Coproduction

Coproduction is dedicated specifically to the study of an emerging field in music production musicology. It explores the limits of what this field might be, from the workings of a few individuals producing music together in the studio, to vast contributions of whole societies producing popular music.

Taking a wide-ranging approach to examining the field, *Coproduction* looks through multiple formats including essays, interviews, and case studies, with analysis and commentary of coproduction experiences at Abbey Road studios. It does so by examining multiple disciplines from social science and coproduction in mental health, to philosophy and mathematics. At its extremes (which is the extreme middle and not the blunt 'cutting edge') the authors attempt to produce every song in their development of an all-encompassing pop music concept, peculiarly called Toast theory.

In attempting to unite the pragmatic collaborative patterns of Vera John-Steiner with philosophical postmodernist concepts of connection, *Coproduction* has something to offer readers interested in the traditional workings of teams of producers, as well as those seeking to understand the wider philosophy of collaboration in music production.

Robert Wilsmore is a composer, producer, musicologist, academic, and collaborator. He studied Music at Bath College of HE (now Bath Spa University) and was awarded Doctor of Musical Arts from Nottingham University in 1994 where he studied composition with Nicholas Sackman. He has led on nationwide research projects on collaboration and has written many articles and chapters on popular music and music production. In his time as an academic leader for more than 20 years, he has been Assistant Head of Music at Leeds College of Music (Leeds Conservatoire) and Head of the School of the Arts at York St John University.

Christopher Johnson is a producer–songwriter and multi-instrumentalist who is perhaps best known in the progressive rock niche for his work with Mostly Autumn, Halo Blind and Fish. He has collaborated on more than 25 studio records, maintains a busy touring schedule, and is a Senior Lecturer on music production courses at York St John University. He is currently working on his PhD, which explores various models of collaboration in music production and how they affect the aesthetic of the resulting music.

Perspectives on Music Production

Series Editors:
Russ Hepworth-Sawyer, *York St John University, UK*
Jay Hodgson, *Western University, Ontario, Canada*
Mark Marrington, *York St John University, UK*

This series collects detailed and experientially informed considerations of record production from a multitude of perspectives, by authors working in a wide array of academic, creative, and professional contexts. We solicit the perspectives of scholars of every disciplinary stripe, alongside recordists and recording musicians themselves, to provide a fully comprehensive analytic point-of-view on each component stage of music production. Each volume in the series thus focuses directly on a distinct stage of music production, from pre-production through recording (audio engineering), mixing, mastering, to marketing and promotions.

Gender in Music Production
Edited by Russ Hepworth-Sawyer, Jay Hodgson, Liesl King and Mark Marrington

Mastering in Music
Edited by Russ Hepworth-Sawyer and Jay Hodgson

Innovation in Music
Future Opportunities
Edited by Russ Hepworth-Sawyer, Justin Paterson, and Rob Toulson

Recording the Classical Guitar
Mark Marrington

The Creative Electronic Music Producer
Thomas Brett

3-D Audio
Edited by Justin Paterson and Hyunkook Lee

Understanding Game Scoring
The Evolution of Compositional Practice for and through Gaming
Mackenzie Enns

Coproduction
Collaboration in Music Production
Robert Wilsmore and Christopher Johnson

For more information about this series, please visit: www.routledge.com/Perspectives-on-Music-Production/book-series/POMP

Coproduction

Collaboration in Music Production

Robert Wilsmore and Christopher Johnson

NEW YORK AND LONDON

Cover image: Simon Piasecki

First published 2022
by Routledge
605 Third Avenue, New York, NY 10158

and by Routledge
4 Park Square, Milton Park, Abingdon, Oxon OX14 4RN

Routledge is an imprint of the Taylor & Francis Group, an informa business

Library of Congress Cataloging-in-Publication Data
Names: Wilsmore, Robert, author. | Johnson, Christopher (Producer-songwriter), author.
Title: Coproduction : collaboration in music production / Robert Wilsmore & Christopher Johnson.
Description: Abingdon, Oxon ; New York, NY : Routledge, 2022. |
Series: Perspectives on music production | Includes bibliographical references and index.
Identifiers: LCCN 2021059663 (print) | LCCN 2021059664 (ebook) |
ISBN 9780815362531 (hardback) | ISBN 9780815362555 (paperback) |
ISBN 9781351111959 (ebook)
Subjects: LCSH: Popular music–Production and direction. |
Sound recordings–Production and direction. |
Composition (Music)–Collaboration. | Popular music–Philosophy and aesthetics.
Classification: LCC ML3470 .W67 2022 (print) |
LCC ML3470 (ebook) | DDC 781.49–dc23
LC record available at https://lccn.loc.gov/2021059663
LC ebook record available at https://lccn.loc.gov/2021059664

ISBN: 978-0-815-36253-1 (hbk)
ISBN: 978-0-815-36255-5 (pbk)
ISBN: 978-1-351-11195-9 (ebk)

DOI: 10.4324/9781351111959

Typeset in Times New Roman
by Newgen Publishing UK

To Cath, Eve and Magnus. Together, in the lockdowns of 2020 and 2021, we went through every known model of collaboration. Mostly successfully.

Robert Wilsmore

To all those musicians, for sharing their stages, studios, and records with me. And to Anna and Daisy, for sharing everything else.

Christopher Johnson

Contents

Figures

Tables

Examples

Contributors

Phillip Brady

Having completed a maths degree at Durham, Phil trained to teach at the University of Nottingham. He now lives in North London and teaches at one of the country's leading state schools. Phil is a keen (if incompetent) musician and before the birth of his two sons enjoyed singing in choirs in London.

Ruth Lambley

Ruth is a PhD researcher at York St John University and is exploring coproduction in mental health research, particularly looking at the role of creative methods in facilitating this. She is also the coordinator of the Converge Evaluation and Research Team (CERT), a group of researchers based at York St John University with lived experience of mental health difficulties. CERT conduct both academic research and evaluations of mental health-related community projects. CERT focus on full coproduction with participants and creative ways of facilitating both evaluation and dissemination.

Preface

Coproduction as an area of study within the wider discipline of music production musicology is still in its infancy and we are pleased to be able to add to this emergent field with this book in the *Perspectives on Music Production* series. As it is a relatively new area, we have, in our use of the term coproduction, thrown the net as wide as we can in order to see what the scope of the field might be. Only the first two parts focus on collaboration in the field of 'traditional' studio-based production, albeit this is by far the largest part of the book and possibly the most useful to the student of music production. After that both the terms 'music production' and 'collaboration' are pushed to include as wide a usage as we can, from the multiple voices working together within the individual, to the whole world in collaboration. In brief, we align with Robert C. Hobbs' proposition that:

> collaboration is, in essence, nothing more or less than influence positively perceived as part of an on-going cultural dialogue.
>
> (Hobbs in McCabe 1984, p.79)

The drivers for this project stem from articles we had written prior to the conception of the book and that are published here for the first time. In particular, the dilemma of maintaining an authentic self as a performer in 'Play One We Know' (Chapter 13), the Toast theory of distributed collaboration (Chapters 14 and 15), and an investigation into how remixing has pervaded music throughout history in 'The Ancient Art of Remixing' (Chapter 16). As we hold a shared interest in how others impact on the Self and vice versa, and as we were already collaborators on compositions, on productions, on chapters on music production, it seemed inevitable that we would write this book.

We are aware, as theory in this particular field is still developing, that we are having to draw from other disciplines such as the social sciences, philosophy, and mathematics, in order to meet our task, and at times we have worked with experts to help fill the gaps in our knowledge and skills (as Bruce Woolley, legendary songwriter with Grace Jones, advised us "Collaborate. Find the people who can do what you can't"). And we are also aware that there are failings to our theoretical frameworks and their application that render the discourse sometimes awkward and naive, sometimes speculative, or even pretentious. But our aim is always to elucidate the practice of our subject at a pragmatic level as well as draw it into the realm of philosophical abstraction, and then to bring these things together where we can, unifying, as Hegel would put it, concept and reality, universality and particularity, understanding and sense. Because this is not easy to do the results can be awkward, but we hope our work here gives something to push off from, or to push against, in the continued development of research into collaboration within music production.

Robert Wilsmore and Christopher Johnson

Acknowledgements

With thanks

To our colleagues and series editors Russ Hepworth-Sawyer, Mark Marrington, and Jay Hodgson for trusting us with this project as part of the *Perspectives on Music Production* series and to Ben Burrows and all at York St John University for their support. To those in the mental health projects that we worked with on both Abbey Road case studies included here, Chris Sheehan at Karousel Music, Paul Pritchard at Abbey Road, Richard Clarke at Monk's Road, Alex Mann at Help Musicians UK and everyone involved on the *Smoke Rainbows – Music Minds Matter* album. To Converge and the Communitas choir of York St John University, in particular Nick Rowe, Esther Clare Griffiths, Chris Bartram, Ruth Lambley for her chapter, and our coproducer at Converge Faith Benson, for their work on the 'Nothing of Value' project. To Ed Coulden, Jonnie Khan and all at LS18 Rocks (*we gotta keep on singing!*). To Bryan Josh, Olivia Sparnenn-Josh, Iain Jennings, Angela Gordon, and everybody with Mostly Autumn for all the years of musical collaboration. To Halo Blind for their patience in trying out different ways of producing together. To Phil Brady, although the BWW label contains three names the system is almost entirely of his making, and to Neil Ward whose programming skills are making the system a reality. To Dale Perkins and Leeds Conservatoire and to Adam Stanovic and Hull University for inviting and encouraging the presentation of the papers that would become Toast theory. To Michael Ward at Leeds Beckett University and to Phil Harding for his incredible openness and generosity. To Bruce Woolley for the insight into studio collaboration and to David Young for the line of flight with the Theremins. To Simon Piasecki for his enduring support and incredible artwork for the book cover. To Abigail Hall, Joe Collins, and Angus Williams for their vocals in The And and 'And And And' and '*The Song of a Thousand Songs* Song'. To Alison Rigg for proofreading and to our editors. To Matthew Reason for his continued advice and calmness under pressure, to Vanessa Corby for the Planet Rock critical listening sessions, and to Gary Peters for deciding to play the pedal steel guitar rather than write about Hegel for us.

Credits

Cover artwork by Simon Piasecki
Photography in Figure 4.2 by Chris Readman
Photography in Figures 9.2 and 9.3 by Esme Mai
(all other photographs, figures etc. are by the authors).
'Ampersand Clef' logo designed by Robert Wilsmore and Robert Oldfield.

How to Read This Book

Design and Structure

Many of the ideas and approaches to this book owe a great deal to the postmodernism of Deleuze and Guattari, and although it may seem outdated to be taking lines of flight from their *thousand plateaus* forged in the 1980s, we feel that these concepts still have much to offer in furthering our understanding of collaboration and music production. With this in mind, the authors of this book on coproduction are not necessarily expecting the reader to tackle this cover to cover. Rather, we *expect* a 'dip in and out' approach; the student looking for a quote for an essay, the reader only reading the bits that mention Brian Eno, the academic looking to use Toast theory or to criticise it (you're most welcome), but we are not expecting many to tackle this cover to cover. Why should you? We didn't write it that way.

If particular parts of this book appeal, then read them, or re-read them, and skip the bits that are of less interest. The translator of *A Thousand Plateaus*, and philosopher in his own right, Brian Massumi analogises that book to the 'record' in his translator's foreword, making it an especially fitting comment for our book too. He writes:

> How should *A Thousand Plateaus* be played? When you buy a record there are always cuts that leave you cold. You skip them. You don't approach a record as a closed book that you have to take or leave. Other cuts you may listen to over and over again. They follow you. You find yourself humming them under your breath as you go about your daily business.
>
> (Massumi in Deleuze and Guattari 1987, pp.ix–x)

At this moment in time the analogy would be that of the playlist on a streaming service rather than the record, and the ability to cut out the tracks is even easier today, we simply have to say out loud "Skip" to get to the next song. The book format, in print at least, has to start at page one, but we don't have to read from page one. Some of it might make more sense if read in order. Otherwise, for example, the reader skipping to the end may wonder why we are attempting to write 'every tune' as a mathematical sequence in a book on music production if they have not read Toast theory first, and the latter itself might seem similarly out of place without understanding the distributed pattern of collaboration set out at the start. A linear reading may have some benefits, but it is certainly not a prerequisite to approaching this book.

Simon Piasecki's Escher-esque book cover artwork captures our themes and our approach. There is no right way up, we appear multiple times coming and going pulled by different gravities, fruit is as prevalent as musical instruments, figures behind the glass are tangential rather than central, and lines and staircases are slanting and oblique.

Book Structure: The Four Parts of the Book and Why

Given the Deleuzoguattarian influence, why have a structure to the parts and chapters at all? There is a second and equally strong influence at play in this book, that of the collaborative patterns of the Vygotskian social theorist Vera John-Steiner. Her four distinct (but overlapping) patterns of collaboration give us the framework to focus each of the parts on. We have adapted them for our own purposes so that they have a certain coherence of thought relating to the areas we wish to explore, and we are aware that this is at odds with the Deleuzoguattarian rhizomatic approach, applying borders and ends where none should be permissible, but it gives the reader and the authors something to hang on to.

The four parts are:

1. Group coproduction (collaboration between individuals)
2. Internal coproduction (the self as many)
3. Coproduction without consent (denial or unknowing collaboration)
4. Deproduction (the collective disappearance of production)

These types are detailed in the introductory chapter 'A Typology of Collaborative Practices in Music Production' (which is an update from a previously published chapter in this book series titled 'Towards a Typology of Collaborative Practices in Music Production'). Here we provide just a brief summary of them.

Part One 'Group Coproduction' focuses on studying producers and teams of producers and artists that work together in the traditional sense of artists working on producing a song or an album. We look at the different modes of collaboration that Vera John-Steiner sets out within these groups, as well as using her patterns, in a modified way, to govern the larger structure of the part.

Part Two 'Internal Coproduction' focuses on the individual producer and the production environment (the Production Habitus) that comes to bear on individual choice. Drawing on Bourdieu's notion of habitus it explores the norms of the environment, its hierarchies, its implicit rules, and the battles within these, particularly where other creative artists and audiences are involved. This part also concerns the internal voices of others in one's artistic decisions and addresses the authenticity and identity of the producer and performer that arises from these, often conflicting, voices.

Part Three 'Coproduction Without Consent' explores the notion that everyone involved in music, creators and listeners alike, are collaborating on the production of the pop project. It is 'distributed collaboration' taken to its extreme, and for us the extreme is in the middle rather than at the avant-garde cutting edge, which seems blunt to us now. The key ideas for this are captured in the concept of Toast theory, which has nothing to do with lightly burnt bread but stands for 'The Song of a Thousand Songs', a reference to the biblical *Song of Songs* and the Deleuzoguattarian *Thousand Plateaus* (with a passing nod to Campbell's *The Hero with a Thousand Faces*) and considers all songs to be a part of one big song that we are all writing.

Part Four 'Deproduction' extends the idea of Toast to logical, if somewhat bizarre, eventualities where society as a whole has done away with production. It is speculative but we have also tried to make a start, one rather small step towards doing this by beginning to set out a logical way of writing every tune. Just as a Chinese proverb has it that a journey of a thousand miles begins with a single step, so 'The Song of a Thousand Songs' starts with a single (83 million tunes long) sequence. We do worry though that the fun part of this journey will be over by the year 2100, and that we will be seeking elsewhere for thrills whilst music production declines to nothing.

The overall construct is this: we start with investigating the traditional practice of producers collaborating together, then we zoom in to the individual and the close surroundings that act on them, then we zoom out as far as we can to see the whole world as collaborating on one big song, then we zoom forward to see where it might all end (or not). If you are looking for guidance on which bit to read, then:

- If you are interested in production teams working together, read Part One.
- If you are interested in authenticity, identity and how the producer is affected by their environment, read Part Two.
- If you are interested in postmodern philosophy and music, read Part Three.
- If you are interested in speculative thoughts about the future of music production, or just like getting annoyed with pretentious ideas, read Part Four.

The Voices of the Authors

The book is not a collection of loosely connected essays, but it has been planned around existing, mostly unpublished, articles and ideas (as we noted in the preface) and forged into a coherent body of explorations on the subject of coproduction. And neither is it of one voice, chapters have been given the names of the authors; we wrote in parallel or we wrote in sequence, rarely have we written 'jointly' in the manner that Deleuze and Guattari did. Where we needed an expert from another discipline we have drafted them in to write separately or to contribute a complementary skill in a co-authored manner. The two main writers are myself, Robert Wilsmore, and Christopher Johnson. We are both composers, producers, performers and academics but also quite different in our approaches and our interests. Perhaps being older I am the one who clings on to the fanciful ideas of postmodernism, having lived through the shift from modernism to the postmodern, and where my musical education was largely classical and which privileged modernism. Cage and Birtwistle were, and are, amongst my heroes, but yet I feel privileged (similar to series editor Mark Marrington, I think) in regard to being excited to listen to Stockhausen one moment and Beyoncé the next. There is something in this book about trying to understand what we went through as musicians in the latter stages of the twentieth century, caught between those that thought there was a high and a low culture and the 'omnivorous' that didn't. KLF's Jimmy Cauty and Bill Drummond, in the guise of The Justified Ancients of MuMu, wrote *The Manual* not just to inform others how to have a number one record but also "to try and understand the whole process ourselves, make sense, unravel the mess of confusing strands" (2001, p.143). We have tried to do some unravelling, and we have tried to proceed with the spirit of playfulness that The JAMs exuded. We are curious. We wonder what The JAMs would have to say about Toast theory.

Whereas Part One has both of us writing for it, chapters in Part Two are entirely by Chris and hence they are his voice and chapters in Part Three are entirely by myself, reflecting our interests and approaches, our writing styles, ideological differences and lived experiences. Ruth Lambley's voice is her own and comes from a discipline outside of music and production. Mathematician Phillip Brady did all the maths for 'writing every song' and wrote much of the explanatory text that accompanies it.

There should be books that focus just on the traditional teams of producers and how they collaborate, like Howard Massey's *Behind the Glass* but with interviews of teams rather than individuals. Perhaps these now already exist or will do soon, but that is not this book, this is a quirky, interdisciplinary study on coproduction that reflects the interests of the authors and the opportunities that have come our way in the process. It is an addition to the emergent field of research into coproduction, and one that we hope will help articulate your own thoughts more clearly and lead to further interest and research in this area.

1 A Typology of Collaborative Practices in Music Production

Robert Wilsmore

Intro: The Dominance of the One

The music producer has long been represented by the image of the lone figure of a man sat at a large mixing desk strewn with knobs and faders. All that has been captured from the virtuosic performers, creative artists and genius songwriters is now at his fingertips with faders moving up or down at the inspired will of this solitary guru. The tracks laid down by the artists are just the raw materials to which this great sculptor of sound will roll up his sleeves and mould the music into the fixed-for-all-eternity artwork that is the recording.

It is a romantic and seductive image, albeit the 'his' stands out uncomfortably now and the image of the man at the desk fits the recording engineer more specifically than the producer who might not even touch a console in the making of a recording. Nonetheless, the 'one man and his desk' image persists. But we do not always think of the producer as one person in isolation, we may sometimes think of them as a member of a team, of a band, but not a member to whom we would generally attach the angst and expressive rawness of the artists who make the sounds. Although some were as dramatic and temperamental as the artists they produced, the producer is often the one who saves the artist from their failings and cuts the rough diamond into the shape that gives the sound its aesthetic (and increased monetary) value. George Martin was often thought of as one of the Beatles, not the first Beatle but the *fifth*, the 'quintessential', Beatle, and this recognises his importance as being part of the music, placing him as the calm, in-control but nearly out-of-sight, mastermind behind the sound of the records. And perhaps this is no illusion, as far as any truth can be true (which is a tough call in this day and age), we rightly revere those who have sat behind the desk on the other side of the glass and pushed and pulled the faders and routed this auxiliary through that effect. There were and are many great controllers of sound who can lay genuine claim to this singular title of 'the producer'. The lone figure is not a myth, it is just that it is not the whole story.

This singularity is further made evident in the literature of producers. The reading lists for our students of music production are likely to include classic works such as Howard Massey's *Behind the Glass* (2000) with more than 30 chapters, each of which is titled only by the producer's name, it features interviews with "top record producers" such as Brian Wilson, George Martin, and Nile Rodgers. And I make no criticism of this, as for any budding producer here are the 'great' producers telling us how they made the works that we know and love, and beyond that, Massey brings out the human in these people. They are not gods born with superhuman powers capable of doing things we will never be able to achieve, not at all. As the front cover says, this is "how they craft the hits", for indeed it is a craft, and the producer does not necessarily appear *deus ex machina* at the end to make the problems go away and bring a glorious final resolution to it all. I say 'not always' because there is probably enough evidence to suggest that at times, they are literally the god from, or with, the machine, and perhaps it is the machine that gets in the way of our understanding of producers and

DOI: 10.4324/9781351111959-1

production. The machine in this case is not the "plain boxes" described by Eisenberg that sit in our living rooms and "can play any damned thing we please" (2005, p.187) but a placeholder for the machinery of the producer, the most predominant of which is the console.

Anthony Savona's *Console Confessions* (2005), from the same Backbeat Books publications as *Behind the Glass*, although different in approach to Massey's book, still has lists of those great names attached to the chapters: Les Paul, Phil Ramone, Herbie Hancock and so on, and, like Massey, the book has those iconic images of the man and his machine – the console. In both these books, the bearded and bespectacled Ramone looks directly into the camera lens with the roomful of gear as the background landscape, like a music studio take on a Reynolds or Gainsborough portrait but with knobs instead of trees. In another picture, Herbie Hancock relaxes, hands behind his head, smiling at the camera off-centre to the left of shot so that the many faders and lights are visible to the right. In a similar position, in Massey's book, sat relaxed with arms behind his head but with even more faders than Hancock, Humberto Gatica smiles (more cheekily than Hancock) and the shot is on a diagonal that maximises the landscape aspect's capability of fitting in as much of the desk as possible across the diagonal from bottom left to top right. To be fair, George Martin is appropriately sat behind scores and a pencil and not a console, and Geoff Emerick is sat smiling on a sofa that is evidently not in a control room. Not every representation defaults to the iconography of the desk, but it is still the dominant image. Head to the last section 'Part Six: Young Guns' and you will once again find the image in every portrait.

But these great books are intentionally focused on or around the desk; it is, usually, the thing directly 'behind the glass' and it is in the title *'Console' Confessions*. I am not criticising these brilliant and insightful books, merely raising the issue that the power of the iconography of the desk may be reductionist to the exclusion of the wider richness of the producer's art. The console is of course an impressive instrument, and we are amazed that this relaxed and in-control expert knows what all these buttons and faders do. Most people would probably be clueless as to what to do if they came face to face with such a beast. It looks like rocket science, and these guys are creative engineers capable of getting a spacecraft to the moon and back. Yet they sit there and smile as if it's all in a day's work for them – which of course it is. The images are incredibly seductive, and they are, in one sense at least, genuine. The literature and the recordings testify to their ability to use the equipment as the tools of their trade, so why not have Jack Douglas lovingly caressing the faders or Frank Filipetti sat at the console, hands folded holding an unlit cigar? Is it any different to picturing Jimi Hendrix with a Fender Stratocaster or Jimmy Page with a Gibson Les Paul? The answer is "Yes, it is different"; hence, there is a mythology at the route of the producer's art that has been oversimplified in the iconography. To say that Jimi Hendrix played the guitar on 'Purple Haze' is probably about as accurate as we might get with an attribution of what the artist did. He played the guitar, and we can hear it. The contribution is easily identifiable. We can make out the guitar's sound quite clearly and attribute it (quite clearly) to one man. The guitarist is the person who plays the guitar; the producer is the person who produces. The obvious discrepancy is that where playing the guitar is clearly understood as the activity taking place, the act of producing, as we know, is multifaceted with no clear boundaries as to what it includes and excludes, and because we (the fans, the interested) want clarity, then the desk becomes the catch-all placeholder signifying activity well beyond what the desk itself is capable of. Hence it plays into the fixity of signification that we associate with the guitar, but in reality, try as the image might, it lacks the guitar's distinct denotation and by attempting to represent something that has no such definition, it enters into the realms of dubious connotation and myth.

The iconography of the man at the desk as described here is a rough generalisation, one that we will instantly recognise whilst at the same time we may have a nagging doubt (or even be quite sure) that the music producer extends far beyond this simple, yet dominant, representation. In *Organizing Genius* (1997) Bennis and Biederman wrote in the first chapter 'The End of the Great Man':

> But even as the lone hero continues to gallop through our imaginations, shattering obstacles with silver bullets, leaping tall buildings in a single bound, we know there is an alternative reality. Throughout history, groups of people, often without conscious design, have successfully blended individual and collective effort to create something new and wonderful.
>
> (Bennis and Biederman 1997, p.2)

The ideology of the producer (but not the reality) is in some ways akin to the notion of a lone hero such as the classical composer. They too are, or were, almost exclusively male and singular in their authorship; whereas we can quickly go straight to Lennon and McCartney when asked to give an example of a song writing duo, we are likely to struggle when asked to do so for classical composers. That symphony co-written by Beethoven and Mozart never happened, and that opera by Verdi and Puccini never happened either. Even the idea sounds odd but why should that be? Songwriters collaborate all the time, yet somehow 'composers' don't, or at least they only collaborate in complementary mode with other disciplines, such as with poets and librettists. But in contrast, production is not without its great teams. In the '80s, we might well have been used to the names Trevor Horn, Giorgio Moroder, and Quincy Jones, but then another name on everyone's lips at that time (in the UK at least) was SAW (Stock Aitken Waterman) with whom we associated the great '80s names of Kylie Minogue, Rick Astley, and Bananarama, among so many others, with whom they developed ways of turning out hits that decades later are still loved, admired, and respected. And maybe this has something to do with the team spirit and the collaborative processes out of which they were forged. A spirit that knowingly drew inspiration from the Motown model of collaboration. A key figure at PWL studios (the "factory floor" of SAW), Phil Harding, writes openly about the creativity of collaboration with his colleague Ian Curnow, writing with wonderful generosity that:

> There's something unique about working partnerships where you get to the point where you know what each other are doing and thinking without verbal communication. In other words, a telepathic understanding where you get on with creating what you feel is right.
>
> (Harding 2010, p.129)

The term *production* is immediately thrown wide when we speak of Stock, Aitken, Waterman, or of Motown before that. Hepworth-Sawyer and Golding in *What Is Music Production*? (2011) note that the three general pillars of production are "capture, arrangement and performance" and each of these has many sub-categories as well. We have long accepted that the production line involves stages that might otherwise be named differently, such as composition, performance, engineering, arranging, or even inspiring (definitely 'inspiring', getting the best from one's performer), and it is hard to give it a definition; least to say that usually there is a 'recorded' element in there somewhere. It is easy for us to understand the idea of one person as the artist or as the producer and of the individual who has a role within the team. But what is *joint* authorship? What is coproduction?

Rather than attempting an ontology, the aim here is to find the different ways in which coproduction occurs. It would be nice to find that moment, one clear example, where we can with absolute certainty say "there is a singular joint authored moment", not a moment where we can identify individual contributions or where we cannot identify individual contributions only because we cannot see the history of its production, but because this moment is genuinely a result of "when two become one". In *Behind the Glass*, Massey transcribes this conversation with Brian Wilson:

> [Massey] Recently, though, you've been working with coproducers as opposed to being the sole producer. Is there a reason for that?
>
> [Wilson] It's because I needed to have the springboard of ideas between people. Because I'd run out of ideas – I had writer's and producer's block.
>
> (Massey 2000, p.43)

There are many reasons for coproduction – the sparking of ideas, the complementarity of skills, the desire to work together, the potential for greater success of the whole through team effort. And there are many forms that coproduction can take; from two producers knowingly working together to a whole world that works together on a singular project (either in denial or without realising it). Even the individual may have to acknowledge that they are also others in themselves, those internal voices of one's influences. The aim here is to forefront joint authorship in producing music, production that is coproduction, that is multi-authored in all its manifestations, from the agreed collaboration of two producers working closely together, to the many and varied contributions to a project by contributors who have no such contract or contact with those making the work. Even solitary gurus cannot override all the decisions that have been made leading up to the point that has presented the material before them; production has already taken place, and they have to work with that as part of their product. This chapter attempts to put into types various modes of coproduction that we will explore in greater detail later.

Type Overview

Only part of the book will focus on those groups of producers who work together to produce their product in the traditional sense, that is, the team that is attributed to the finished recording of the song. Although that study deserves greater research in itself, to begin with, we want to throw the net as wide as possible and find not only the cutting edges but the *extreme middle*, a place that was sadly neglected in a modernist last century. We are under no illusion that the main interest will be in the teams of producers and how they work together, the actual people coproducing actual hits, but in this first attempt, we want to zoom out to view as much of the territory as possible, even if that means covering ground that some will find uninteresting and possibly unconvincing. This is not then a study just of those collections of individuals who form production teams, albeit that category and the types within it are part of what is outlined here, and we recognise that most interest is likely to lie in that area rather than in our more fanciful philosophical speculations and assertions that occur in later parts. There are different modes of collaborative practice that we can bring to this aspect of the study, in particular we have chosen those identified by Vygotskian collaborative theorist Vera John-Steiner, but we are also excited to venture into less conventional areas of coproduction and take a look at the individual as a collective in themselves, as well as the individual's place within a collective of actual others. Then we observe that music in general, and popular music in particular, is an ongoing coproduction by the many millions of contributors who are involved in producing this ongoing singular thing (popular music) that we have called 'The Song of a Thousand Songs' (or 'Toast' for short), which is something of an 'all the world's a stage' approach to pop music collaboration. We are aware that not everyone in coproduction is a willing collaborator and acknowledge that plagiarism is also a form of joint production (if collaboration can only be called such if all participants are willing and in agreement, if consent is a predicate of collaboration, then we are at fault in our project). We also wonder where it could all end up; this is in the realm of philosophical speculation, but we will look to a future point where production may have ceased because all has been collectively produced. Not a Hegelian 'End of Art', where art's function or use (to the *Idea*) has ceased but where there is actually nothing left to produce. We will even attempt to do that ourselves by exploring how all musical phrases might be systematically produced and categorised. We will, without any doubt, fail but perhaps we can put the wheels in motion and in doing so we might spark a few thoughts and ruffle a few feathers. The feathers of authorship and ownership in music production could do with a good ruffling.

To put these larger categories into manageable types, we list the *top-level* categories of coproduction as:

1. Group coproduction (collaboration between individuals)
2. Internal coproduction (the self as many)
3. Coproduction without consent (denial or unknowing collaboration)
4. Deproduction (the collective disappearance of production)

Each of these will be introduced separately, mostly with further subdivisions within categories. Some of these types are drawn from observation of the real world, of an engagement with the act of production and coproduction, but some of the models are also set up *a priori* so that they may be tested against real situations later. Hence the types will shift from the obvious to the ludicrous (by which I mean "playful" in its exploration of ideas rather than demanding universal validity). Ultimately, it might turn out that coproduction is nothing more than a discipline-specific study of teamwork. If that is the case, it will still be an interesting area to study if not a paradigm shift into some exciting new realm of music production studies. That remains to be seen.

1. Group Coproduction

Emerging from a Vygotskian social science approach to joint authorship, Vera John-Steiner in her seminal book *Creative Collaboration* (John-Steiner 2000) identifies four "patterns of collaboration" that she carefully caveats as being on a fluid spectrum, and that these move from the closest of collaborations to the widest and most open form (and vice versa). They are analysed as Integrative, Family, Complementary, and Distributed. These have associated roles, values, and working methods, and they form a useful model for understanding different types of coproduction. John-Steiner's model, neatly depicted as a circle in order to avoid a hierarchical taxonomy, starts from the widest heading and goes to the most integrated; however, we will work outward only because the most open collaboration (distributed) leads to our later types, that of 'coproduction without consent' and 'deproduction'.

1.1 Integrative Coproduction

This is the mode in which we might best hope to find that purest of coproduced moments. The integration of producers at this point is such that their understanding has moved beyond the need for discussion, beyond the need to divide roles, in fact beyond any recognition as to who contributed what to the point where there is no difference between the operation of the one and the operation of the many. This moment may not actually exist outside of theory, even in a hypothetical example it seems hard to see how there could be this pure moment. One might imagine two producers with hands on the same fader making adjustments almost as if receiving spiritual guidance on a Ouija board, but even then (ghostly externality aside), we are drawn to the separate decisions that provided the pressures on the fader and hence are thrown back into knowing that there is separation. Perhaps integrative coproduction is in *proximity* of this pure moment rather than achieving it. John-Steiner's descriptions of this type are of "braided roles", "visionary commitment", and "transformative co-construction" (2000, p.197), the braiding confirming that separate strands still exist making this model far more discoverable than the pure moment might be. That said, Keith Sawyer, in his research on 'group genius', points to moments of group decisions that are exactly this, that is, where the group finds a solution to a problem but none of the individuals within the group are aware of that solution. Following a narrowly avoided shipping disaster in 1985 on the USS *Palau*, researcher Ed Hutchins analysed the responses and actions of those involved in averting the disaster and concluded that "the solution was clearly discovered by the organization itself [the collective group] before it was discovered by any of the participants" (Sawyer 2007, p.28). So, there may yet be hope of finding similar moments of coproduction that are outside that of any individual contribution.

1.2 Familial Coproduction

Trust and common vision are central in this type. Coproducers will be highly familiar with each other and will have become used to each other's ways of working and cross over roles easily. John-Steiner's term is 'family', but we have used 'familial' here to emphasise that the mode is 'typical of a family' just to avoid any misunderstanding. In this mode, any division of labour might change, and expertise becomes dynamic. This is a pre-integrative model where there are still identifiable contributions but much crossover. Here, one might find coproducers feeding in and out ideas and activity in all stages of capture, arrangement, and performance, comfortable enough to know that one has permission (indeed encouragement) to do so from the 'family'. One producer has gone into the live room to move the kick drum microphone that the other set up without asking because they know they can act upon their own informed decision to do so and that consent is implicit (Eno and Lanois had such an agreement).

1.3 Complementary Coproduction

Perhaps the most common form of coproduction is the mode in which roles are clearly identified and executed as such. In this way, one can assemble a team of experts trusted to do their activity best, and it will benefit the whole. One producer might be able to EQ the singer's voice better than the other, but the other has inspired the singer to give the best performance in the studio and so on. Separation of roles is key here but so is an "overlapping of values" (John-Steiner 2000, p.197) so that the final product works as one unified thing.

The artists themselves are perhaps the most frequent coproducers, and most likely, this operation will fall into the 'complementary coproduction' type, that is where the division of roles is clear and complementary, but there is a shared vision (if it is to be successful) and the artists will often be with the 'producer' validating their decisions or trying out ideas with them in the control room. Complementarity is also probably the best understood of collaborative types. Where John-Steiner uses 'complementary', interdisciplinary collaborators have used the Deleuzian terms 'striated' and 'stratified' (see Alix, Dobson, and Wilsmore 2010). Joe Bennett, in song writing collaboration, uses the term 'demarcation' (Bennett 2011), and business teamwork theorists have this complementary model worked out in great detail, in particular the roles identified by Meredith Belbin that produce successful working teams (see, for example, Belbin.com).

1.4 Distributed Coproduction

In terms of 'normal' studio practice, this would seem to be a somewhat anarchic working model, as its methods are "spontaneous and responsive" with roles that are "informal and voluntary" and might even lead to sabotage of others' contributions. This model suggests a rather more open approach that might see the production group widen to anyone who might want to 'have a go'. Hence, this is not likely to be a typical model in terms of song production in a studio. We will need to look wider to see this model in operation. Artists sometimes put out stems for fans to mix and produce as they will, and this fits this model well. Moving out even further, we might view the artist who coproduces by using a sample of a previous work, with or without consent. Further out still, we may see the whole of popular music as one production that everyone contributes to, and indeed that notion will be explored in more detail under the discussion of 'coproduction without consent' (Toast theory) which aligns with John-Steiner's role description of "informal and voluntary" collaboration.

2. Internal Coproduction

As with the patterns of collaboration, the types identified here will often overlap, or they may even be considered as the same model but viewed from a different angle. And so it is with internal coproduction. We might recognise that our decisions that ultimately manifest themselves in terms of what comes out of the studio monitors will include thoughts of those who *have* influenced us (past) and thoughts of those who *will* listen to us (future). Unlike the composer who walks a dangerous legal tightrope when stringing together a sequence of notes, the producer has been freer to take on influences of the past with less fear, though even that territory is now beginning to be claimed and fenced off. Within the self, one might have to identify one's influences *and* one's audience (past and future others). The discussion relates to artists of any type and any discipline with regard to decisions one might identify as one's authentic self and then those decisions that second-guess what the other wants (the audience). Heidegger notes that the inauthentic self includes the other as our "*they-self*" in that "The Self of everyday Dasein is the *they-self*, which we distinguish from *authentic Self* – that is, from the Self which has taken hold of in its own way" (Heidegger 1967, p.167). For Deleuze and Guattari, the "I" does not disappear but rather becomes much reduced. As they write with regard to their collaboration on *A Thousand Plateaus*, they wanted "to reach the point, not where one no longer says I, but the point where it is no longer of any importance whether one says I" (Deleuze and Guattari 1987, pp.3–4). Not only are they collaborating with each other; they also acknowledge that as individuals, they are already a collective of others. We can try out two models in this category, one that is before one's production and one that considers the "after" of one's production.

2.1 Present–Past Internal Coproduction

The influence of existing production comes to bear on one's own production. To produce with producers of the past, one does so presumably without their consent (which also means this sits in another category), but if one is to stand on the shoulders of a giant, does one need to ask the giant's permission to do so? Although one could, it is not every case that one *has to* return to one's 'collaborator' for permission after making a contribution. Perhaps this type does not need a new name; we can call it 'influence'. It is simply that of the self as constructed from the many, but it gives an opportunity to think of our engagement with the past as a collaborative venture. The past is tried and tested; hence, we might go straight for that Neumann U87 for the vocals and an AKG D12 for the kick drum because it has worked *before* for others. Or perhaps we can be inspired by Martin's discussions with Lennon that gave rise to the extraordinary 'Strawberry Fields' recording where the desire to achieve something new required a problem-solving approach to production – the past telling the future to try out new ideas. In this particular example, Lennon liked the first part of one studio recording and the second part of another and wanted to combine them. George Martin pointed out that they were in different keys and had different tempi, to which Lennon replied, "Yeah, but you can do something about it, I know. You can fix it, George" (Martin and Hornsby 1979, p.200). And, of course, he did (*deus ex machina*).

2.2 Present–Future Internal Coproduction

The future of one's production lies with the listener. In this mode, we are familiar with what they will like and give it to them, or we give them something we know they will like but that they don't know yet (a disruptive model in the manner of Henry Ford or Steve Jobs). Either way, although the second notion seems more creative, both involve the consideration of the judgement of the other that is yet to come. The wanting to succeed, the need to be wanted, the need for money and so on drive an inclusion of the future *other* in the production decision. Do we have to tune that vocal? Whom are you doing that for? Is it because we believe the audience demands it? Another *other* here are the artists themselves of course to whom (one version of) the producer must be subservient or simply be one voice in the crowd, but this model best sits in the complementary coproduction model. Richard Burgess' 'type C' producer, the collaborator, has similar properties to this (see Burgess 2002). Here we are in a dual position of basing future audience expectation on past audience behaviour or basing future demand on our intuition that goes against past audience behaviour (they haven't behaved like this before because they've never heard it this way before). The latter model has been the dominant model of creativity, certainly in modernist times that privileged cutting ties to the past, but we are less convinced that it works now. It is time to retreat from the cutting edge (have we not noticed that it has become blunt?) and move instead toward the middle (toward James Blunt) for that is where the creativity of this age really lies.

3. Coproduction Without Consent

Here we will outline two modes of coproduction without consent, although we will also have to say up front that there is also a 'knowing' in this category as well as an unknowing; however, as a general heading, it will work for the moment. The first type is that of the distributed model of collaboration, as discussed earlier, but in this case it has been widened out to encompass the whole world, so that everyone who produces popular music is a collaborator on one singular song, which we introduced earlier as 'The Song of a Thousand Songs'. The second type in this category is that of denial or unknowing collaboration. In this type, it might indeed be 'knowing coproduction' but the person denies this knowledge, where an artist has deliberately taken the work of another and used it in their own creation, though more often than not, this category will not be aware that it is coproducing but rather will be asserting originality and authorship. This category has clear connections with that of the 'internal coproduction' model described earlier, as it draws externality into an internal creative process.

3.1 The Song of a Thousand Songs (Toast Theory)

This type is actually where this whole project started, having its roots in a paper given at the International Festival for Artistic Innovation at Leeds College of Music in March

2014, which was called "The Song of a Thousand Songs: Popular Music as a Distributed Model of Collaboration" (IFAI 2014), and bit by bit the theory has expanded to encompass the many other modes that are now being explored. The notion of joint production, unknown (or unthought about) collaboration where everyone in the world is continually adding to this singular song that is called 'The Song of a Thousand Songs'. Theoretically, we have found that the articulation of this concept works best as a combination of the social science distributed model of collaboration by John-Steiner and of the postmodernist ever-connecting rhizome of Deleuze and Guattari. In this way, the extravagant and poetic philosophy of postmodernism is held to account by a grounded social science approach or at least by a large dose of pragmatism. And they are not mutually exclusive; in fact, the multiplicity of the rhizome is simply another version of distributed collaboration, de-centred and de-ontologised in order to point out the singularity and connectedness of things rather than the separations that may indeed not be there at all. Since its inception, the title itself 'The Song of a Thousand Songs' has become rather too cumbersome for frequent use and the written abbreviation of 'Tsoats' was soon corrupted to become 'Toast', and hence 'Toast theory' is the generally used term that stands for 'The Song of a Thousand Songs'. In the book on Mixing in this series on Perspectives of Music Production, we discussed the issue of the ontology of the mix in 'The mix is, the mix is not' (Wilsmore and Johnson, 2017) that the separation of songs in our everyday world is nothing more than a prescribed division, where these separations (ends of songs) seem to have nothing at all to do with sound and as such should not be thought of as endings. We wrote that:

> If songs are separated largely by non-musical signifiers (the composers, the song titles, etc.) we can "nullify endings and beginnings" [Deleuze and Guattari 1987, p.25] for these starts and finishes are nothing more than segmentation caused by the effect of imposing non-audio signifiers onto audio. When we do this, we cease to operate within a representational system. There are no longer identifiable ones of songs – the removal of the artificial beginnings and ends has shown that they are actually all joined together. In fact, it is not correct to say that they are joined at all, once we have removed the sticky labels marked 'beginning' and 'end' we see underneath that there is nothing but continuity.
>
> (Wilsmore and Johnson 2017, p.196)

In one respect, the model is far-fetched and a long way from our everyday understanding that individual songs are composed by individual artists or bands, but then it is also a familiar notion that music has a life of its own and that "rock 'n' roll ain't gonna die", even if that is only as a metaphor rather than a belief in music as a living thing. And the main focus here is to point out the grounded observation that endings and separations of songs are indeed mostly about something other than the sound itself, and in doing so, we hope to clarify what endings are in the 'real' world (in our everyday phenomenology) and how this might affect considerations of what to produce and what coproduction means. This postmodernist take is not intended to be sophistry, it is intended to help show how the world actually works in 'real' life.

3.2 Denial or Unknowing Coproduction

There is a familiar saying that "where there's a hit, there's a writ", and we never have to wait very long to read in the press another case of artists (or their estate or their lawyers) laying accusations at the feet of another artist who they believe has stolen their creation. At the time of writing, there was a particularly interesting one that the press reported. It is interesting because the song in question has already been through the legal process of another artist laying claim to it itself and the song having to state that it was co-authored already (at first, it was coproduced without consent, but then consent was retroactively applied). In this case, BBC music reporter Mark Savage, reported the following headline, "Lana Del Ray Says That Radiohead Are Suing Her" (BBC 2018), and on her Twitter account, she let it be known that:

> It's true about the lawsuit. Although I know my song wasn't inspired by Creep, Radiohead feel it was and want 100% of the publishing – I offered up to 40 over the last few months but they will only accept 100. Their lawyers have been relentless, so we will deal with it in court.
>
> (Lana Del Rey 2018)

The song in question is Lana Del Rey's 'Get Free' (2017). The BBC article then notes that "Interestingly, Radiohead themselves were successfully sued by The Hollies over Creep's similarities to 'The Air That I Breathe'. Albert Hammond and Mike Hazlewood are now listed as co-writers for the song, and split royalties with the band" (BBC 2018). Larry Bartleet writing for the *NME* gave some clarity on the issue:

> Radiohead's publisher denies that any such lawsuit exists but explains that they are asking for Radiohead to be credited on the song. A statement from Warner/Chappell reads: "As Radiohead's music publisher, it's true that we've been in discussions since August of last year with Lana Del Rey's representatives. It's clear that the verses of 'Get Free' use musical elements found in the verses of 'Creep', and we've requested that this be acknowledged in favour of all writers of 'Creep'. To set the record straight, no lawsuit has been issued and Radiohead have not said they 'will only accept 100%' of the publishing of 'Get Free'".
>
> (Bartleet 2018)

Not surprisingly, those on social media responded in a variety of ways in support of, or against, each artist, but many noted the problem of owning a melodic sequence or a chord structure. So if Lana Del Rey's song 'Get Free' is adjudged to have come from 'Creep', and 'Creep' from 'The Air that I Breathe', then we can trace that song back to the one before it and that song to the one before that, and so on, until we get (hypothetically) to the very first song. We have done the "research" and have found the *very first* song, the song from which all others come, and the very first song that was written before all other songs is called 'This Song Sounds Like Another Song'. If Warner/Chappell want an acknowledgment in favour of "all writers of 'Creep'", then it should hold that every song credits every contributor from the writer(s) of 'This Song Sounds Like Another Song' onwards. Ridiculous, of course, but which bit is ridiculous, my example or the real situation?

Baudrillardian notions of simulacra in recording are a firm part of the music production curriculum now for those of us teaching music production musicology in higher education, and Auslander, pulling on the work of Gracyk and Baudrillard, notes the simulacra of the recording as the copy that precedes the real (see Auslander 1999; Baudrillard 1994 [1981]; Gracyk 1996). The concept behind this hypothetical first song 'This Song Sounds Like Another Song', is of course to highlight the questionable link between origination and ownership, as cosmologist Carl Sagan said, "If you wish to make apple-pie from scratch, you must first create the Universe" (Cox and Ince 2017, p.78). This system of ownership will have to crash at some point in the future, but for the moment at least it is fair to say that the Lana Del Rey and Radiohead conflict will not be the last of its type and the crash will not be happening anytime soon. And so, lots of songs (that sound like other songs) have been claimed to be owned, and the lawyers set to work. So, we can legally own those things as writers and we can lay claim to putting one note after another in a particular sequence and own it (and defend it in court), so can we do the same with creative production? What if I creatively EQ, fix a specific compression ratio, and pan the guitars hard left and right? Can I own this? Maybe it is no different except that it can't be policed as well (yet) or maybe it's that no one cares as much. I should add that we are not advocating a 'copyleft' ideology that "seems to argue for the reduction of, or extinguishment, of copyright altogether for an alleged greater cultural good" (Bargfrede 2017, p.xi) or indeed are we affirming the copyright; rather, we are exploring the discourse. Perhaps, to retain our Deleuzian leaning, we should consider ourselves as the copy '*milieu*' or 'copymiddle'. The extreme middle, the distance furthest away from the now blunt cutting edge, is our favoured territory.

4. Deproduction

At a hypothetical point in the future, when collectively we have produced all that can be produced, the territory of production will begin to vanish. The process toward this is that of deproduction, and it might be said that this process has already begun. In reality, the final absolute point is unlikely ever to be reached, but the process toward that end will have a significant impact upon how we view production both by reframing the local and ultimately by accelerating output so that, if the totality of production is finite, output stops (the coal has been mined, the pits closed and dismantled, the term 'mining' no longer in use or understood).

The rise in technology that has driven the decentred Internet and globalisation has produced some fine examples of distributed models of collaboration that Vera John-Steiner might never have considered even just two decades ago. Where there were territories that had borders and where maps remained fixed (at least for a while), there was some sense of security, but that has been replaced by the security of the omnipresent, that is, our ability to access information and experience anywhere, anytime. The authenticity that decided that the singer was only authentic if their singing accent represented their hometown (the physical territory) has given way to something much more local, namely, the global. The voice in one's own room (in the city of York, England, for example) that is the recorded voice emanating from the speakers, might be from New York but that voice is *nearer* to the listener physically than the next-door neighbour and is more present in the house than the neighbour is. In this case, the territory, the social locality, has been reterritorialised by the global. Arguably, when a singer is in a Yorkshire studio in the flesh (the one from York but with the Brooklyn accent), we can confidently say that their accent is authentic, because they *relate* to the voice that is *nearer* to them more than the singer from York with the Yorkshire accent. That singer has heard the NY accent more in their own home than they have heard a Yorkshire accent.

Wikipedia is a good example of the global over the international and of a distributed model of collaboration. In one entry on deterritorialisation, including a discussion on Anthony Giddens's notions of globalisation from *The Consequences of Modernity* (Giddens 1991), there is a particularly revealing intervention; the entry includes an edit in superscript in square brackets from a contributor who, in doing so, wittily demonstrates exactly how a distributed model operates:

> In the context of cultural globalization, some [*weasel words*] argue deterritorialization is a Cultural feature developed by the "mediatization, migration, and commodification

> which characterize globalized modernity". This implies that by people working towards closer involvement with the whole of the world, and works towards lessening the gap with one another, one may be widening the gap with what is physically close to them.
> (Wikipedia 2018)

The global intervenes by becoming closer in proximity than the local (it is nearer to you than your neighbour). The permissible intervention (the contribution of the phrase "weasel words") likewise destabilises the local ownership of knowledge, just as it destabilises and then re-stabilises the authenticity of the recorded voice.

As the world is at liberty to collaborate, to contribute, and to produce, and technology increases the rate and quantity of contributions, we can consider what might result should it turn out that production is not infinite but finite. To transfer this to music production, can we reach a point where we have collectively produced every 'recording' that can be produced to the extent that not only does further production become impossible but the *Idea* (and its apparatus) also falls out of use and hence out of existence? How could that happen? Here is one small way to start: Write every tune that ever did, and ever could, exist. Impossible perhaps, but we can have a go. We can begin step by step. Most tunes fall within certain parameters, maybe a few bars, normally within a key or mode that is limited; normally, it will be a tempered scale that is fixed, note lengths might be between a whole note and a sixteenth note, and so on. Suddenly, limiting parameters emerge that make it look slightly less than impossible, achievable even. And they could be written systematically and categorised, particularly if we can write the software to generate these. We would encompass every tune (at least within the chosen parameters) that has been written, and we would also then have written every tune that is to come. What then of ownership? Perhaps the money we will owe, because we will be sued for writing tunes that are already 'owned', could be offset by the income from those future composers that we will sue because we have already written the tune that they will claim as their own. Answering the same concern from readers of *Wired* on his article "How many different songs can there be?" (Wired 2015), Rhett Allain responded similarly, writing "If I sue more than I get sued I should be fine." Just imagine how that might play out in a court of law. As mentioned earlier, at some point, ownership will have to come crashing down, and that will be one of the stages on the journey of deproduction. It does not need to reach the final realisation of all production for the process to have impact. Beyond this, we are not claiming a 'death of art'. Music will be there, in fact, *all* music will be there already, it is just that production will have stopped.

Outro: The Rising of the Many

This last stage is conjecture and speculation, but it is not without actual purpose. Somewhere along the way to these concepts there lie a few shifts in our collective understanding of music production and of authorship. And as we accept our collective approach to production and let go of the individual as producer, then we may find that we are freer to produce. Coproduction, in its reduction of the "I", ought also to reduce the grip on ownership that stifles creative acts and creative generosity. It may turn out that the field of study of coproduction is nothing more than how the known types of collaboration are manifest in the discipline of music production. If so, then that is fine, it is still worth the effort. And we do not lay claim to this territory, we are neither the first nor will be the last to research it. We are just another local calibration in the wider rhizome. However, it is time for the rhizome to spread, coproduction may yet have much to offer as an activity and as a field of study.

Part One

Type 1. Group Coproduction

Collaboration Between Individuals

2 Producing Together

Robert Wilsmore

Introduction

The producer is a mysterious being; it is hard enough to know what one of them does let alone two or more of them producing together. Production is an ambiguous and catch-all term that does not lend itself to transparency. As Hepworth-Sawyer and Hodgson put it, the work of the producer can encompass "everything done to create a recording of music" (2017, p.xii); as such the term is a facade behind which many varied activities may have taken place and the facade is, at best, opaque. From the outside we are, as the biblical metaphor has it, "looking through a glass darkly", and we look through a muddied window squinting for a glimpse into the control room. Here we are back at the 'producer as god' analogy and the *deus ex machina*, the god and their machine, the workings of which are shrouded in mystery to the general observer. But there is of course a thrill in the mystique of what is going on, an imagined alchemy that creates gold from base metal and it might spoil our excitement to know how mundane the creative process might actually have been. David Howard in *Sonic Alchemy* writes a short account of a collaborative venture between Brian Wilson and Gary Usher on a ballad called 'The Lonely Sea' in 1961 in Southern California:

> That first night the two collaborated on their first song together, a beautiful dreamy ballad called "The Lonely Sea". Wilson decided he wanted the sound of real waves on the song and convinced Usher to help him lug the cumbersome Wollensak recorder to the ocean in the middle of the night.
>
> (Howard 2004, p.52)

In this anecdote, taken at face value, the co-labour that takes place involves the moving of a heavy tape recorder from one place to another, and the story continues with Usher's effort to find someone else who would plug it in (a third 'collaborator'). The *Idea* appears to be Wilson's, so does he merely need a hand with the carrying? Perhaps that is all this amounts to and that this is enough to label it a collaboration; the work of a producer and a recording engineer perhaps. We might venture though that there is more to it, not just the wider events that might have happened that night but also the intricate and undocumented exchanges between the two that made the event possible and began the journey towards a shared vision. One may need the others' courage as well as their muscle, their complicity and their validation. Note too that Howard recounts that Wilson 'convinced' Usher to carry the recorder and there is one school of thought that considers collaboration to be nothing more than 'influence'. And of course, this anecdote is more than a carrier and an 'ideas' person, this is a *story*, another romantic image to thrill us and add to the mythology of the record and the producers, the mundanity is juxtaposed with the romance of the timing of "the middle of the night". We may fixate on the studio, on the technology, but so much of production occurs outside of these, instead operating inside of relationships or networks,

DOI: 10.4324/9781351111959-3

operating through 'influence' where relationships can be taken to involve human and non-human actors (things that 'act' on other things), to use a term from Actor–Network Theory (ANT) led primarily by Bruno Latour (see, for example, Latour 2007, 2017), where a flat ontology does not accept the *a priori* dominance of the human over the non-human. We have found in our studies, perhaps to the reader's disappointment, that we spend little time focusing on "who pushed which buttons" but rather choose to ask the question "who pushed whose buttons"? If, as Robert C. Hobbs claims, "collaboration is, in essence, nothing more or less than influence" (McCabe 1984, p.79), then we are in search of influences. As composer Dimitri Tiomkin, blurring the lines between influence and collaboration, said on receiving an Oscar in 1954 "I want to thank my collaborators, Bach, Beethoven, Brahms and Debussy" (Hughes 1979). We may influence others, be influenced by others, or be under the influence of alcohol, drugs, technology, or maybe even fruit.

In our own observations whilst recording at Abbey Road we often discuss the Abbey Road fruit that sits in a bowl on the table up in the control room of Studio 2. Far from the cigarettes and alcohol iconography of studio behaviour that we see in documentaries, the fruit suggests an alternative ideology, one that acknowledges that the art of recording has matured since the heady days of excess in the studio, a serious art rather than a bohemian passion where the birth of the great record is at the expense of the well-being of the suffering artists. The Abbey Road fruit became a talking point, almost becoming part of our conscience sitting on our shoulders, albeit in the shape of a banana rather than an angel. It aligns with the professionalism of the studio engineers we worked with; it is a non-human actor that plays its part in the network of 'actants' in the production process. We are not suggesting that we were collaborating with fruit, albeit that would be a logical conclusion if influence is collaboration and fruit has an influence, only that every small part of the environment (the sofa, the scented soap in the toilets) acts upon the environment and in turn upon the collaboration, and in considering the questions "who pushed whose buttons?" or maybe "*what* pushed whose buttons?", we should remember that someone *chose* to put that fruit there, someone chose the sofa, and someone selected the fragrant soap for the toilet. It does not need to be that the soap was selected with the *intent* of it playing a part in the success of the recording but that it *may* play a part. The environment acts upon the production and hence the product. The 'structuring structure' of the habitus (Bourdieu 2010, p.166) nudges and pushes actions in particular directions, and actor network theory asks us to note *all* things that act upon an eventuality. We noted in our article 'The Mix Is. The Mix Is Not' (Wilsmore and Johnson 2017), that when the verb (mixing) ceases to be an action and stops, it becomes the noun (the mix); in this case, the act of producing, when action stops, becomes the product, and the 'producing' contains so much more than is generally acknowledged. As Brett Lashua and Paul Thompson noted:

> Quotidian creative practices in studios are often left uncritically examined. Much like microphoning, comping is largely absent from popular representations of studio practices. Even the slightest activity can be complicit in the production of recording studio myths.
>
> (2016, p.85)

There is, most likely, more intent in 'comping' and 'miking' than choices of soap with regards to the recording, nonetheless even the very deliberate procedures of comping and miking skills are often overlooked, so what chance do soap and fruit have? And yet, if it is fair to say that if "even the slightest activity is complicit" then we must consider all actants within the habitus of the production. Flatten the landscape, lay low the peaks, we will see more that way.

Returning to Brian Wilson and Gary Usher's 'The Lonely Sea', the weight of the Wollensak, the ocean, midnight, and both Wilson and Usher are all actors in the wider production. As mentioned in the opening chapter, Wilson said in his interview with Howard Massey that he needed coproducers "because he had run out of ideas". But that was much later than the Usher collaboration on 'The Lonely Sea' when he presumably did have ideas. Complementarity in this model (artist and labourer) is less equal (when traditional notions of hierarchy are reintroduced), less of a shared effort to produce the whole but rather how one person needs to draw on others with Wilson as the primary actor to which other heterogeneous elements "do their bidding" (Michaels 2017, p.154).

To continue with the mundane, pragmatics often drive the need to work with others, for one Abbey Road album recording session in 1969 George Martin had orchestral players in different rooms:

> In the first studio Martin conducted one set of musicians, while Geoff Emerick, Phil McDonald, and Alan Parsons supervised the other portion in the second room. The two teams then communicated back and forth via walkie-talkies to synchronise the recordings.
>
> (Howard 2004, pp.34–35)

In this example the collaborative mode is one of cooperation and coordination, that is they had to 'work together' in order to get the timing right. Engineers solving a problem. Again, like the carrying of the recorder, this is an operational decision, yet it is still something that ultimately is about the resulting sound, and hence no surprise that the creative break between engineering and production was crumbling even then. These examples are mundane, they are not of the level of sorcery that is the myth of production, yet they also speak of creativity and where the settings are far from mundane (the beach, Abbey Road studios, etc.). There are two extremes at play here, the image of the producer–gods that move in mysterious ways, and the image of the producer–labourers lumbering a heavy tape recorder towards a beach. We may even have to relent that, after the glass has been cleaned and we can see more clearly, that there is no mystery, no magic, no absolute integration at all, just joint labour, some fruit, and somewhere to plug in the speakers.

Patterns of Collaboration

We have set out in the introductory chapter our version of Vera John-Steiner's four patterns of collaboration (Integrative, Familial, Complementary, Distributive) as the benchmark for our first general type of coproduction, that of group coproduction (collaboration between individuals). In this section we will outline how other typologies in the field of production, in particular those of Edward Kealy and Richard Burgess, map against John-Steiner's patterns before then giving further examples of our four types.

The Wilson and Usher example demonstrates the characteristics of complementary collaboration where there are clear divisions of role but with 'overlapping values'. That relationship may have become familial or even integrated at a later stage, but 'The Lonely Sea' example taken on its own, and that it is their 'first song', is one of complementarity. This particular division of role is explored and extended by Edward Kealy in his article 'From Craft to Art: The Case of Sound Mixers and Popular Music' (see Kealy in Frith and Goodwin 1990) where he focuses on the position that the 'sound mixer' or 'recording engineer' (he uses both terms) has with regard to aesthetic contribution to the recording, recognising this as a collaborative act, and noting that it was not until later in the 60s that sound mixers began "aggressively asserting the aesthetic importance of their work" (1990, p.217). Here, with the shift from 'craft to art' we see the beginnings not only of a move from one to the other but of the *integration* of those roles, the culmination of which is the producer–composer, which is the focus of Virgil Moorefield's book *The Producer as Composer* (2005), representing a new point that we have arrived at several years since the time of which Kealy writes, and it is clear from his account that such roles were complementary rather than integrated until the latter half of the twentieth century. Kealy puts forward two modes of collaboration in record production, namely the 'Craft-Union mode' and the 'entrepreneurial mode', and he is confident and rightly unapologetic in his use of the terms collaboration and collaborator, given his essay is about the creative contribution of the engineer it is correct, and in doing so helps to break the rift in status between engineer and artist. The craft–union mode of the '50s is characterised by a division of labour enforced by union regulations. Here the record producer is ascribed both management roles (contracts, time management) and artistic roles (selecting and arranging music) but no mention of a mixing desk; that was (by union regulations) the sound mixer's domain, and the sound mixer was one of the collaborators that the 'record producer' coordinated. He writes:

> The company designated an administrative supervisor to recording sessions, the artists-and-repertoire man or "record producer," whose duties included expediting compliance with the contractual provisions of the collaborators, coordinating their work, keeping the studio sessions within budget and on schedule, and selecting and arranging music to suit the company's intended audience. [...] The relationship among collaborators at such recording sessions tended to be formal and impersonal.
>
> (Kealy in Frith and Goodwin 1990, p.211)

But the media was changing in the '50s. With the rise of television addressing popular culture and radio addressing localised tastes, and with the lowering of costs of studio equipment, entrepreneurs were able to set up smaller operations, and with this they could operate outside of the union restrictions of larger corporations:

> In contrast to the craft-union mode, with its emphasis on technical correctness, concert hall realism, and strict division of labour, the entrepreneurial mode is a more fluid and open collaboration which allows an interchange of skills and ideas among the musicians, technicians, and music market entrepreneurs. [...] this integration of functions also had important consequences for the sound mixer. In exchange for the opportunity to contribute to shaping the musical aesthetic, he also had to share his control over, and knowledge of, the studio technology with his collaborators.
>
> (1990, p.213)

The description aligns Kealy's entrepreneurial mode with that of 'familial coproduction'. In John-Steiner's patterns the 'family' pattern of collaboration is characterised as 'fluidity of roles', the values of 'common vision and trust', and the working methods of 'dynamic integration of expertise'. This also maps to Kealy's descriptions of 'fluid and open collaboration', and of 'dynamic integration of expertise', although it seems slightly reluctantly with regards to the sound engineer having to hand over a "share of his control" in exchange for aesthetic input. It was the small companies that operated nearer to the idea of a family than that of the large corporations whose visions were with regard to technical excellence and reproducing the concert hall with high fidelity. Theirs was an elitist vision, it need not be a *shared* vision by all employees who have jobs according to union guidelines. It might be true that the small, familial companies need not have a shared vision, but there is more likely to be greater buy-in to the vision, even if that is relatively short lived, as in the case of Stock

Aitken Waterman, and where, like many bands themselves, the intensity can create growing rifts that can bring an end to the group.

George Martin similarly discusses the separation and the crossover of roles noting that:

> A producer's function is to listen to the sound, and to the music as an overall unit together, and from that he must judge the recording. An engineer's function is to ensure that, technically, it is the very best recording obtainable. [...] Equally, there can at times be a legitimate overlapping of function.
>
> (Martin and Hornsby 1979, pp.249–250)

Albin Zak notes that "the engineer is, in a sense, a translator for the other members of the recording team" (Zak 2001, p.165). Like Healy, Martin recognised a change over time from where the division of roles was almost a "matter for the union" to the position where the end result, the success of the recording, drove the activity and roles rather than the job description.

Producer and writer Richard Burgess presented his typology of four types of producer in his questioning chapter 'What kind of producer do you want to be?' (Burgess 2002). Type A, 'The-all-singing-all-dancing-king-of-the-heap', does everything themselves and avoids the input of others, but he cautions that "Working by yourself can sometimes get a little stale, so it's better to have a collaborator around to help keep up the inspiration level" (2002, p.4). A point that is affirmed in Brian Wilson's comment on why he worked with coproducers. Type B 'The faithful sidekick' gets something of a drubbing as "No-one ever wants to own up to this stereotype", and somewhat ironically for this book, they are "almost invariably credited as co-producer" (2002, p.5). The sidekick, amongst other things, is relegated to taking on jobs that the main producer or artist is less keen on doing. Burgess rallies to support this role writing "the title may sound demeaning, but the position is definitely not" (2002, p.5). The collaborative pattern is complementary in this type, possibly with familial aspects in the manner in which family groups can have dominant members. If our book has an aim beyond merely observing, researching and philosophising about coproduction, then that aim is to move on from the demeaning position from Burgess' observation and help give the title of coproducer the weight it deserves, so that the title and the position can carry equal authority.

Burgess' Type C the 'Collaborator' has close links to the characteristics of familial coproduction as well as those of complementary coproduction. They will have often been in a band themselves before moving to a producer role and as such "They have most likely always enjoyed collaborative situations and they bring that band-member-mentality to their productions" and their tendency is to steer the group towards a group decision rather than to take decisions unilaterally. However, the collaborator steps out of the band and facilitates *their* sound and vision. Michael Jarret's chapter 'The Self-effacing Producer: Absence Summons Presence' (in Frith and Zagorski-Thomas 2012), documenting mostly jazz and country producers, writes of most of these producers that "They are self-effacing. They believe the producer who serves best is least audible or intrusive or noticeable" (2012, p.129). This collaborator is a team player, not someone who needs to make visible their own mark. For our purposes Burgess' description of the collaborator is too narrow, given our task is to explore collaboration in its most extreme forms (such as in the Toast theory) and its most mundane (carrying a tape recorder or putting fruit on the table). Jarret's self-effacing producer fits with both the band-mentality of Burgess and the complementary of John-Steiner.

Type D is 'Merlin the Magician' who is an "intangible force" who commands loyalty from record label and artist alike but may not spend much time in the studio at all, "In a way he acts like a hands-on A&R consultant, coming in with an objective view frequently referred to as 'fresh ears'" (Burgess 2013, p.10). Interestingly Burgess' study has similarity with Kealy's account of the '50s 'artists-and-repertoire man' or 'record producer' which highlights the multiple ways in which the term 'producer' has changed since then. Jarret noted that "The wide-spread adoption of magnetic tape in the mid-1950s demarcated a line between A&R and production" (in Frith and Zagorski-Thomas 2012, p.145). So how would Merlin fit our four types? Merlin is unlikely to sit comfortably in distributed coproduction, although they might want to add their magic to a project if they are so inclined (but they would want the credit). They can operate in complementary mode so long as they are in charge and everyone else plays their roles 'beneath' them; they might not wish to let go of the control needed for familial mode, but potentially, if they could find someone of equal genius with whom they could work with that could produce the 'transformative' vision that the integrative mode is characterised by then they might be part of an integrative coproduction model. John-Steiner's example of this transformative type is that of Braque and Picasso jointly creating cubism, it is a type of collaboration that creates magic. Merlin might be capable of such feats with another like-minded Merlin, egos permitting.

We see then, through the examples above from Kealy and Burgess, with reference to John-Steiner, Jarret, Martin, Moorefield, and Zak, that production roles have shifted over the course of more than half a century from productions that had delineated and complementary roles, to the multiplicity of types of joint authored productions that we are now observing. Like any typology, our types are only a way of observing characteristics and behaviours. We have set out our four types in this category below with examples. No example fits neatly into any one category and most will show characteristics from other types as well. Models are not used to pigeonhole but for comparison, and through comparison to give insight into the subject, collaborative practice in this case.

Integrative Coproduction

Amongst the poker-faced 'geniuses' of hidden producer–gods were the pioneering electronic duo behind Kraftwerk, Florian Schneider and Ralf Hütter. It is hard enough when watching Kraftwerk perform 'live' to work out who is doing what. I query 'live' only in as much the biggest cheer at one gig I attended (Sheffield, UK, 2017) was when the actual band are replaced by their robots and a recording of 'The Robots' is played. A perfect Auslanderian moment of confronting what it means to perform 'live'. For Kraftwerk this mode of performance is perhaps less of a deliberate obfuscation but rather that their authentic mode of performance is that of the man–machine hybrid and not the authenticity of the guitarist, who by hitting the guitar strings harder makes the sound louder. The latter gives the rock guitarist a direct connection to their sound production and with that the permission to raise their arm high in the air before bringing it down fast on the strings. That isn't going to work for a band that only needs to press a button or turn a knob to make it louder. One may pound on a piano to genuine effect but it would be somewhat ridiculous to do so on a harpsichord that does not change its velocity, hence Kraftwerk are rather more Bach than Beethoven in their performance mode (although more Beethovenian when it comes to the thematic construction of their music). But with this lack of outward expression comes a sense of closedness, it exemplifies the difference between the ability to describe, as we noted in the introductory chapter, what the guitarist does (play the guitar) much more clearly than what the producer does (produce). The mode in which the creation of expression occurs is not immediately disclosed in Kraftwerk either in performance or in the studio if indeed there is a distinction. As Moorefield writes in *The Producer as Composer*:

> After *Autobahn* [1974], Kraftwerk began to produce themselves on a permanent basis. They became known as denizens of the studio, mysterious entities who created their purely electronic music behind closed doors.
>
> (Moorefield 2005, p.90)

Perhaps we need to further delineate our integrative model into 'pure integration' and 'perceived integration', the difference being that the latter has no *visible* sole authorship but may have 'actual' individual contributions. Whereas 'pure' integration has no individual contributions, visible or not, rather like the sea rescue example noted by Keith Sawyer in Chapter 1, where the group found the solution and not any one individual.

One revealing chapter on collaborative practices in the literature field is that of Albin Zak's writing on the complementarity of 'Engineers and Producers' in his book *The Poetics of Rock* (2001), and he runs into the same problem of trying to uncover what is going on as Moorefield noted of Kraftwerk. He pulls on a different analogy to express the notion of the 'goings on' behind closed doors, quoting Jerry Wexler he writes "[I]t's like who does what to whom in bed. Nobody knows ..." and goes on to note that it "focuses on the creative roles of engineers and producers, two members of the team whose efforts remain, for the most part, behind the scenes" (Zak 2001, p.164); as such, Healy's writing plays a significant part in Zak's study also, giving us a clearer view of what did happen.

This mystery is intriguing. We have both the analogy of the wizards behind the curtain using their extraordinary powers of sorcery to produce gold (Kraftwerk), and the analogy of sexual activity in the bedroom (Wexler), the actual goings on being the knowledge of the participants only. If we are only seeing the beautiful lovechild that resulted from the activity (the recording), rather than seeing the activity itself, then our imaginations may run riot filling in the missing knowledge gap and hence attributing the greatness of the record to the lover's passion or to the sorcery of wizards, when in fact it might well be mundane events that lead to the product. Perhaps notions of integrative and familial coproduction are less a reality but are rather something that we *attribute* to the producers because we fill in the gaps in our knowledge of what actually occurs. Pure integration may be a romantic idyll only and there is only complementarity when all the details are available to us. But it should not be a contradiction that creativity may also be a mundane activity, as the saying goes, "1% inspiration, 99% perspiration".

Perceived integration may occur where past behaviour enables collaborators to predict how contributors will proceed in a future event. As Phil Harding describes in our interview with regard to SAW (see Chapter 6), conversation on repeated activity is no longer necessary once practice is established:

> Within a production and writing team it's quite incredible how little needs to be said, and the conversations tend to be based around the problem area rather than what needs to be done. In other words, if one party is expecting the other party to do a specific thing and it's all sounding great or it's quickly editable there is no discussion needed.

Discussion is minimised to where the event is outside of normative practices and doxa prevails within the Production Habitus. Fred Seddon, drawing on Roslyn Arnold's concept of 'empathic intelligence' and 'empathic creativity' (Arnold 2003), puts forward his notion of 'empathic attunement' that he observes in the moment of performance for jazz musicians. Like Keith Sawyer (2007) and Gary Peters (2009), the improvisational element of jazz in its moment of *being* provides an excellent place to study and observe how musicians collaborate, how they make decisions that are both on the spot *and* prefabricated. How quickly can the performer see what is being offered, then without time for significant reflection turn to their own internal library, their 'stock on the shelf' or 'tissues of quotations' (Barthes), or maybe invent something not in stock, and return it, in some cases instantaneously with that 'knowingness' of what the other is about to put forward. Seddon notes the shift from sympathetic attunement to empathic attunement as "in order to reach empathic attunement musicians must decentre and see things from the other musician's point of view" (Seddon in Mielle and Littleton 2004, p.68). This attunement between musicians "although rooted in the sharing of stocks of musical knowledge evolves beyond this process. It is proposed that empathic attunement is synonymous with 'striking a groove'" (2004, p.69). And when musicians are empathically attuned, through a process of non-verbal collaboration they "seemed to respond to each other in an atmosphere of risk taking and challenge which extended their knowledge base" (2004, p.73). In our interviews for this book we noted this with producer and songwriter Bruce Woolley (see Chapter 8), collaborating with Grace Jones, where creation of the song was spontaneous in the studio, often working into the night, the process of composing and producing coming out of a moment where a rhythm meets agreement with both of them, and the presence of the drum machine, a TR808 (a non-human actant) was key to the spontaneous development of the song ('Party Girl', in this case). Here the ideas appear to be simultaneous in the manner in which Seddon notes empathic attunement in jazz musicians, and in the case of Woolley and Jones, the composition–production process is simultaneous, given the simultaneity of the recording and composing processes. To return to SAW, as we noted in the introduction, Phil Harding and Ian Curnow shared "a telepathic understanding where you get on with creating what you feel is right" (Harding 2010, p.129). As Keith Sawyer notes "Psychologists call these shared understandings *tacit knowledge* – and because it's unspoken, people often don't realize why they are able to communicate" (Sawyer 2007, p.51). In this respect, integration aligns with our second type of coproduction, that of 'Internal coproduction (the self as many)'. Harding is so familiar with Curnow's decision-making that his own decisions have effectively become a 'joint' decision. In collaborative theory Charles Green (2001), citing the practices of artists such as Gilbert and George, Abramović and Ulay, concludes that there is a 'third hand' at play in such cases. The third hand, although a strong candidate for pure integration, does not necessarily exclude perceived integration (identifiable but unseen individual contributions) as its generative mode.

Being attuned to others is key to being influential in production. Martin noted that the most important attribute was in fact 'tact'.

> Tact is the *sine qua non* of being a record producer. One has to tread a fine line between, on the one hand, submitting to an artist's every whim, and on the other, throwing one's weight about. [...] One had to lead rather than drive. I think that now, as then, is probably the most important quality needed in a producer.
>
> (Martin and Hornsby 1979, pp.44–45)

He follows this with a further attribute, "Another, lesser, quality was the ability to hold one's liquor" (1979, p.45), as this was necessary when dealing with certain artists. Martin's approach to attunement has, in the case of "holding one's liquor", a slightly cynical edge to it, an underhand psychology that is about 'nudging' the artist towards the desired behaviour rather than demanding it. Collaboration in this mode is one of subtle influence applied in order to affect the behaviour of others (finding how to push someone's buttons). There is no doubt that jazz musicians in performance will similarly intend to influence each other, most often this will be open and honest, but sometimes through various tricks to entice the other down an alley of their choosing. It might simply be that the production version of empathic attunement works in a slower timeframe than that of jazz in performance. And who said collaboration had to be above board, fair and equal anyway? Most of the time it isn't. In a research study into how we teach collaboration in universities, Christopher Newell wrote:

> Perhaps the time has come to value the chaos and unfairness of collaborative practice, the cut and thrust, give and take, the absence of reliable rules, standards or processes. [...] The point about collaboration is you may get hurt.
>
> (Newell in Alix, Dobson and Wilsmore 2010, p.57)

And unfair it most certainly is, as Martin frequently refers to, he was often on the receiving end of not very much when it came to remuneration for his contributions to the collaboration. He notes, more with a sigh than in anger, that in 1963 when The Beatles music was being heard in "millions and millions of homes around the world [...] I was still earning less than £3000 a year with EMI" (Martin and Hornsby 1979, p.166). And then another anecdote has it that Martin was paid very little for an orchestral arrangement and to add insult to injury he remarked to his manager Ken East, that "I get fifteen pounds for doing that arrangement. Do you mean to say I've got to pay blasted copyright out of my fifteen quid?" (1979 p.194). This is an essential problem and frustration with collaborative practices, fundamentally the resulting product may be seen as something other than the sum of its parts, but the attribution of contributors, at least when it comes to money, is apportioned in a way that attempts to recognise the parts but ultimately almost always fails to do so. In our interview with Phil Harding in his time with Stock Aitken Waterman he notes the system of 'points' that lead to financial gains on record sales, "At that time I knew that even half a point of a production royalty of something big could be quite a lot of money. For me that would have been a big reward". Generally, one point means 1 per cent with regards to allocating royalties. The term is used similarly in both music and film industries. In this respect being a named coproducer is not about an activity but about money and power, that is, the term is held, and withheld, by those with the power to do so. One becomes a 'coproducer' only if one is *given* the title by those with the power to do so in these cases.

Returning briefly to subtle influence, Thaler and Sunstein, in their work on 'nudge theory' in relation to decision-making and well-being include a study by Matthew Salganik and co-authors in 2006 that details their experiment to see how people are influenced by others with regard to their choice of music to download, a topic close to the hearts of music producers. They noted that in downloading music people were more likely to download songs when they could see that others had already downloaded them; "The identical song could be a hit or a failure simply because other people, at the start, were seen to have chosen to download it or not" (Thaler and Sunstein 2008). So, can the producer use this effect in production? Yes, and in the most extreme form of this they can use a 'big' name that has been seen to have sold records before (and therefore is akin to downloading a song because it has been downloaded by others) in order to override acknowledgement of coproduction and retain the claim of sole producer. As with Phil Harding's example, being a collaborator on a production may not be enough to officially be a coproducer depending on the power structure of the situation. Here we have shifted quickly from attunement in the search for integration to exclusion according to power. Money, it seems, is a divisive influence rather than an integrative one when it comes to acknowledging coproduction.

Familial Coproduction

Although familial collaborations do not necessarily require family connections, they can indeed be related, such as with Motown's Brian and Eddie Holland. The turn of the century production team The Matrix (described in further detail in Chapter 5), best known for bringing Avril Lavigne to the forefront of popular music, includes the then wife and husband pairing of Lauren Christy and Graham Edwards along with Scott Spock. One of the key factors of their success was that being three rather than two generally resulted in majority decision-making rather than the 50/50 stalemate that might arise from a partnership just of two. Christy noted of the three that, "We all have things that we're really, really strong at, but we also involve ourselves in all aspects" (Buskin 2006, p.142). John-Steiner's family model has 'trust and common vision' as a characteristic along with a dynamic fluidity of roles. In the working methods of The Matrix the three had strengths in key areas but were also able to cross over into each other's dominant areas with ease.

But familial collaborations need not be between actual family relations, the term is used with regards to certain characteristics. Mick Glossop (Massey 2000, pp.234–235) describes in detail a series of contributions that led to the big 'Townhouse' sound of the drums (which later became the sonic hallmark of Phil Collins' hit 'In the Air Tonight'). He poignantly describes who contributed ideas, performances, evaluations, technical inputs, as well as environmental factors that were beyond their control but the results of which influenced the decisions of the collaborating group. The case is with regard to a recording session for Peter Gabriel's third album (1980), involving producer Steve Lillywhite, engineer Hugh Padgham, Phil Collins, and Peter Gabriel. Gabriel, imposing an artistic restriction, had decided there were to be no cymbals. The room had a stone-slabbed floor that was not (initially) selected for its acoustic properties. Phil Collins was ("probably") on drums and these drums were heard over the talkback system that was connected to a "really vicious compressor" that was heard through the Townhouse SSL desk. Gabriel decided he wanted to use that sound, Padgham reproduced this sound using the SSL channel compressors. Whilst doing this, he played with the gate whilst 'Collins' hit the snare and the resultant cut-off produced the classic sound. Glossop summarises:

> The sound is basically the sound of a drum being hit and the ambience compressed and then, as the ambience tails off, it chops off sharply because the gate's got a short release. it just doesn't work if you play cymbals as part of the rhythm.
>
> (2000, p.235)

This description gives an account of how the output (the distinctive sound) was arrived at collaboratively through a series of planned and unplanned events that involved both human and non-human actors. It is not the result of one person but of contributors to an iterative process of ideas, performance, evaluation and amendment, a trial-and-error process, that led to a result that was accepted and cherished. The parts, the actors, have been identified so this is not a candidate for a pure integration mode of collaboration albeit they appear integrated as a unit in the sound, there are complementary positions (artist, performer, engineer, producer, etc.) and Gabriel has the executive decision in this anecdote, but there is also a 'fluidity of roles' that characterises familial coproduction, as the trial-and-error process is explored between them (the engineer becomes part of developing the sound experiment). The main point of Glossop's use of the anecdote is to note that the imposed restriction of not using cymbals led to the properties of the environment and the technology becoming active players by 'accident', and that this accident had positive creative outcomes that would not have been there but for Gabriel's initial restriction, or indeed if any of the other actors were not present, regardless of whether they had been planned or not. There is no reason why this would not also sit neatly under the 'complementary' model, but it is part of the familial mode of operation that the documented exchanges between contributors suggests an 'in the moment' fluidity that extends beyond the often neatly delineated separations of complementary roles (of the craft union mode for example) where the 'together in the moment' exchange of ideas is less prevalent.

Complementary Coproduction

The division of labour that separates out complementary skills is most famously exemplified through the workings of Motown records founded by Berry Gordy Jr in 1960. Based on local Detroit car manufacturers, the 'Fordist' approach to record production that Gordy took divided the labour up into component parts where each part was 'manufactured' by the best person for that part with Gordy himself as the leader and the unifying vision that everyone else had to adhere to. The components included songwriters, producers, engineers, lead artists, backing musicians (band, backing vocals, brass), etc. As well as Gordy himself ensuring the aims, objectives and values were met, there were the songwriters (such as Holland–Dozier–Holland), artists (Marvin Gaye, The Supremes), producers (Norman Whitfield, 'The Clan'), and the in-house groups that gave a consistency to the Motown sound, as Stuart Cosgrove details:

> The Funk Brothers often improvised powerful backing tracks that were used and re-used by producers and engineers. The Andantes provided choral support and the vocal undercarriage to assist lead singers and a string of accomplished baritone saxophonists, including Andrew 'Mike' Terry, the purveyor of the legendary 'bari-tacks', delivered rasping support to many Motor city hits.
>
> (Cosgrove 2016)

But Cosgrove is keen to point out that, although there are many similarities, there were also differences, not least that where cars are intended to be exact replicas of each other the songs were not, each bearing the marks of the individual creativity of the contributors. And there were also cross-over of roles, for example songwriters would also be producers, as was the case, for example, where Holland–Dozier–Holland are credited as songwriters for the Supremes hit 'Stop! In the Name of Love' (1965) with Brian Holland and Lamont Dozier credited as producers; Norman Whitehead produced 'I Heard It Through the Grapevine' (released by Mavin Gaye on the Tamla Motown label in 1968), which he co-wrote with Barrett Strong; and Diana Ross and the Supremes' hit 'Love Child' (1968) was produced by 'The Clan' of Taylor, Wilson, Sawyer, and Richards, with the addition of Henry Cosby on the song-writing team. But as with so many production companies, where there is a division of labour and each component is vital to the whole, then each part is also a potential weak link. Hence the strikes and disagreements of the car industry similarly affected Motown. The Clan replaced the departed Holland–Dozier–Holland as the in-house writing team, and they themselves were replaced by The Jackson 5 production team, The Corporation. There were teams working within teams, each with their own *modus operandi*, and it is these working methods that we wish to focus on through examples such as PWL and The Matrix production team in later chapters.

Pete Waterman, heavily influenced by Gordy's approach, developed a British version of Motown with Pete Waterman Ltd and PWL Records, which he launched in 1987. In this mode the hierarchy permits the overriding of individual contribution. It is collaborative but with top-down executive decisions. Here is an example of a small detail of this in action in the studio: What does a producer do when the performer is not up to the job or not doing what the producers want from them? One strategy seems to be simply to lie to the performer and tell them that it is them playing but replace their part surreptitiously. Waterman describes a recording session, several years before PWL, at Trident studios for Stiff Records with fellow producers Peter Collins and Pete Hammond (a very 'Pete' heavy production, although the production credit is generally given to Peter Collins) where, in a studio recording with The Belle Stars of the record 'Sign of the Times' (1982) they wanted to use the Linn drum machine instead of the real drummer. Waterman recounts:

> We had to sit the drum machine under the mixing desk playing it over her drumming and every time she stopped, we'd turn it off. If we were a bit late hitting the button, we told her that it was really her drumming, but that there was a delay in the studio.
>
> (Waterman 2000, p.113)

The approach is product focused, not the self-effacing, type C collaborator, whose aim is to serve the artist, but instead to *be* the artist. The band serves the producer in this example (neither does it go unnoticed that, in this anecdote, the producers are the 'men' behind the desk). For Waterman to progress his ambitions that led to the formation of PWL he used his skill in recognising the skills of others. On meeting Chris Britton (Mike Stock) and his collaborator Matt Aitken, he recognised their musicality and realised his way forward for writing great pop songs. He noticed how quickly Stock and Aitken could pick up on the subtle nuances of the chords, the 'actual' sounds of the recording that made all the difference:

> I knew that the way forward for what I wanted to do would be to make simple songs that had those musical variations in the chords. If we disguised the simplicity of great tunes with complex chords, we could make hits. That was the secret.
>
> (Waterman 2000, p.133)

On Mike Stock he said that:

> The best thing about Mike was that he knew what I wanted almost before I could put it into words. I'd come into the studio with simple melodies and simple lyrics and Mike would turn them around and make them work. The beauty of the way that our relationship developed was that all three of us were integral to an extraordinarily smoothly running machine. Right from the beginning we began to understand that when we were on form all three of us could work together almost telepathically.
>
> (Waterman 2000, p.138)

We can note that in this one production company there were at least two subgroups, SAW as one, and Harding and Curnow the other, both using the term 'telepathic' to describe their working relationships. Perhaps not surprisingly then, when elements of these subgroups came together there were moments of magic. At the now infamous Dead or Alive 'You Spin Me Round' recording session, after the band and Aitken and Stock had gone (with several people having nearly come to blows), Waterman and Harding stayed up all night to mix until finally "Phil turned up one of the instruments and instantly it all came together" (2000, p.151). As harrowing as the collaborative process seemed, the combination of friction and telepathy gelled with one final production decision. Although, officially, Harding was the mix engineer and Waterman the producer, it all added up to a sophisticated approach to pop as an art form. SAW were often maligned as formulaic and commercial, but aficionados will perhaps be pleased to know that it was John Peel, the guru of new music whose approval was a validatory seal for those he championed, who was the first to play a SAW record on radio (2000, p.135).

Meeting one of his Motown idols, Lamont Dozier of the Motown song-writing team Holland–Dozier–Holland, said to Waterman that "I only ever wrote one song, I just did it 66 times" (we would of course point out, from our standpoint of Toast theory, that Dozier was not writing the same song 66 times, but rather he made 66 contributions to the one song). PWL, like Motown before them were to stick to a routine and to the 'Kiss' method, 'Keep it Simple, Stupid'. At PWL they used one microphone for their artists regardless, a "Calrec Soundfield Mic" (now known as Soundfield). As Les Sharma, an assistant engineer at PWL, noted:

> The Calrec mic was connected to the Calrec box, and the box was plugged into the SSL desk, those 2 channels were EQ'ed mostly by an insert point by a good quality EQ like Forusrite, or Neve. In that same insert chain would be a good quality DBX 160

> compressor and DBX de-sser. […] The Calrec SSL channels were then rooted to all channels on the SSL desk, so the recording engineer could at any time select a channel from 1 to 48 to record vocals onto.
>
> (Sharma 2014)

Why mess around with different mics when one will do; keep it simple (stupid). Normalising operations meant that conversation was unnecessary. 'Telepathy' and empathic attunement may then have multiple forms, like-minded individuals find themselves in agreement with the others' contributions, but the same effect occurs when familiarity of operation removes the need to communicate. Operational systems become the orthodoxy, the doxa of the field, of the habitus of the studio and production environment.

Distributive Coproduction

Characterised by a 'voluntary and informal' mode of operation, this form sits more comfortably with Toast theory than it does the recording studio. That said, the internet has allowed collaboration of this type to become widespread, and it comes in many forms. We can see examples of artists opening up their material for others to produce. For example, in 2012 Justin Vernon announced the 'Bon Iver Stems Project' (Bevan 2012) where the stems of the second album had been made available for anyone to mix and enter into a competition. The winners received a cash prize of $1000 and their work released on the 'remix project' track list on streaming sites. This is a controlled model with the artist making available fixed stems and the resulting remixes are acknowledged, which, as such, sets it apart from being 'by' Bon Iver. We might even say that the remix album 'complements' the album from which the stems came, and as such consider this also as a complementary model of collaboration.

Our internet age offers many creative entrepreneurs the opportunity to try out distributed collaborative ideas. In one such example, in 2013 Irish entrepreneur John Holland and his development group WholeWorldBand, developed a phone app that became Youdio, which he described as:

> Think of it as a collaborative, multi-track version of YouTube. […] Fans create or join an existing recording session and contribute a vocal, instrumental, visual, or anything they feel like adding. Everyone from famous professionals to complete beginners can work together to create great songs and visuals using just the camera and microphone in their iOS device – and then share their video mixes directly to YouTube, Facebook, Soundcloud, and Twitter.
>
> (Holland 2015)

It ties in with Toast philosophy, but the 'whole world band' for us has a different name, 'The And', who don't produce lots of songs, just the one very big and ongoing one. The world is our app.

The idea of 'the one and the many' extends at least as far back as Plato's Socratic dialogues (Parmenides in particular) c.400 BC, and our own 'Song of a Thousand Songs' idea dates back to around 2010 as a name and an idea, having its first 'official' outing at a conference at Leeds College of Music in 2014. It is the zeitgeist of our time that 'everything connects' where even the triangular structure of Maslow's hierarchy of needs is often amended so that below it, and hence *before* the need for food and water comes the necessity of 'Wi-Fi'. But there are also more traditional types of production that might be described as distributed beyond the voluntary contribution of internet users in the Bon Iver stems model or the WholeWorldBand app. One way of being a producer is to not produce anything at all but to let others do it all (and leave everything to everyone else but the glory and the credit).

Don't claim it as a 'coproduction' acknowledging the input of others but rather claim the whole thing for yourself. This is obviously a nice way to earn a living as a producer but not a nice thing to do to those that put the effort into producing the music. Not least this model requires that the producer has a 'name', effectively a brand, and as such the name that appears as producer is nothing more than an endorsement and maybe even an endorsement unrelated to quality but rather to quantity, i.e. the amount of cash involved (cultural capital converted to material capital). Not surprisingly, the literature on this issue is either presented as 'blurry' anecdotes, or it is anonymised, or it makes vague claims rather than making any definite accusations. This anecdote about producer Curt Boettcher portrays the problems of acknowledging where credit lies:

> When Curt Boettcher walked into Steve Clark's office with his hot new demo in his hands, little did he know he was about to enter a phase of this career during which he would produce his most commercially successful records under the auspices of Steve Clark's Our Productions. Neither did he know that he would never see a penny in royalties for his work on these records. The first thing Steve Clark did with Curt was put him to work with Tommy Roe. The story gets blurry here. Steve Clark is credited as producer of Roe's hits "Sweet Pea" and "Hooray For Hazel" but Boettcher subsequently claimed that it was he himself who produced them.
>
> (Spectropop N.D.)

Howard writes that "Although Steve Clark retained production credits on the songs ['Sweet Pea' and 'Hooray for Hazel'], Boettcher was the real studio composer" (2004, p.68). Boettcher, coming from a background as a musician, singer, and songwriter having been a forming member of The GoldBriars, a folk-infused band in a time just preceding the Beatles invasion on America, might perhaps have been at a disadvantage in claiming production credits given he had so many other labels by which he may be called (singer, songwriter, etc.). Hence what one is labelled previously might well determine what one is continued to be labelled afterwards. Associations stick, and that holds for most labels and not just in production. Whatever the truth was of this particular occasion with regards to Clark and Boettcher, we can at least state that the actuality of the producer is a contested space made contestable by the difficulty in identifying clearly what the role of the producer is. It would seem possible that being the producer can be a *claim* rather than a role or a skill set, etc. It can, as we have seen in the case of Phil Harding's request for 'half a production point', be about who holds the power to 'give' the label of producer or coproducer.

Martin saw so many different roles for the producer and types of producer, not least the type in this category which explores the dubious accolade of the producer who doesn't produce. In one anonymised anecdote he recounts one 'famous' producer who would:

> [...] sit with his feet up in the control room, watching the group in the studio below. Occasionally he would press the mike button to communicate with the studio and say, 'Absolutely fantastic!' [...] That was his entire contribution. Having achieved all this, he would walk away with a hit record!
>
> (Martin and Hornsby 1979, pp.251–252)

It is, on the one hand, wonderfully outrageous, whilst, on the other, blatantly makes use of the effect of the 'seeing through a glass darkly' aspect of the secretive nature of the producers work. The muddied and obscured window analogy suits the traditional studio and the separation between those making the sound and those secreted in the control room who are either busy or pretending to be busy. One cannot get away with pretending to play in the live room, not playing or playing badly will be noticed but it seems that, in this instance at least, one can pretend *behind* the glass. Perhaps no wonder then that coproducers are seldom credited with the role if tradition is in favour of maintaining that singular authorship and where the role is not as delineated as someone who plays. We need only to refer to Auslander's Baudrillardian critique of the Milli Vanilli Best New Artist Grammy award 1989 scandal where the artist's producer "created fresh controversy when he admitted that not only had the duo lip-synched their concerts, they had not even sung on the recording for which they were awarded the Grammy" (Auslander 1999, p.61). That the performers did not perform becomes a scandal that changed legislation on consumer fraud, yet the fact that a producer may not have 'produced' may pass without such concerns. Whomever it was in the anonymous example of Martin's that did most of the production probably had to settle for some self-acknowledgement that they had done their part and acquiesce to the name and the brand that will help to sell the product. Still, this particular category doesn't seem very fair. But, referring again to Newell's quote, the "chaos and unfairness" of collaborative practice is something to be valued, even if, somewhat ironically, it is the value of the contributors that is unfairly miscredited or unacknowledged. A different value then, the value of the singular product rather than the individual values that could be better acknowledged but for the fixity of the hierarchy that pre-existed the recording and was maintained afterwards. But does this sit fairly in the 'distributed' category? It seems neither 'voluntary or informal', and yet the producers (the 'real' ones) are not named as such. Perhaps 'delegated' production would work better, and that it sits closer to the complementary model, or even a further category called 'unfair complementarity'.

Perhaps the most traditional form of 'informal, voluntary, spontaneous and responsive' form of distributed coproduction, where multiple parties attempt influence simultaneously, is outlined briefly in a subsequent chapter, where the beleaguered 'mix engineer' (officially at least) is bombarded by the chaotic nature of 1970s punk. That is, Phil Harding stuck in a studio with members of The Killing Joke shouting in his ears from all sides. Punk as an ideology, and as a response to the technical virtuosity of prog rock as exemplified in the iconic image of Johnny Rotten wearing his 'I hate Pink Floyd' t-shirt, choose to rebel both in performance and in the studio. Distributed group coproduction can be the enthusiastic fan mixing stems from, and returning them to, the internet, and it can be the chaos of punk rock.

Sam Backer in an interview with producer Glenn Lockett, aka SPOT of LA-based SST records (Backer 2018) who helped define punk rock in 1980s America, exemplifies not only his shift of focus from prog rock to punk, but also the similarities of its chaotic freshness to contemporary jazz. In Backer's interview SPOT comments that in the '70s "I was really enamoured with a lot of what I guess you would call progressive rock," he says. "That whole era with concept albums, when jazz fusion started happening, before it turned into bullshit." Then SPOT met guitarist Gregg Glen, founder of hardcore band Blag Flag. Intrigued, SPOT would occasionally jam with him and the rest of Black Flag at their dilapidated practice space. But it wasn't until he saw a riot break out during their performance at an outdoor concert in Manhattan Beach's Polliwog Park that he realised he wanted to produce them. "That show was just so crazy," he recalls. "I said, 'I got to record this band before they get killed.'" When they did record an album *Jealous Again*

> for months it wasn't clear who would be the vocalist on the record. To top it off, despite their inexperience, the members of Black Flag believed that they knew how to record better than SPOT did, making the recording process a constant struggle for control.

This struggle for control can be both the creative gift of, and the chaotic curse of, distributed coproduction. But as with all examples it shares characteristics of other models, and in the case of punk as described in the examples above, punk, through a very unplanned, unintegrated, and distributed way, was most definitely transformative.

Collaboration in General

Collaboration and working with others, often with complementary skills, is usually the key to all productions, and the lonesome 'King of the heap' is perhaps an exception rather than a norm, albeit 'the heap' are contributors to the project. In Massey's second volume of interviews *Behind the Glass Volume 2* (2009) he subtitles the opening chapter with a Nashville producers' panel using the phrase "It's all about the relationships", and upon asking the question "what advice can you offer the young record producer?" Justin Niebank states the turning point for him was when:

> I made the conscious decision to quit chasing the brass ring and focus on the people instead. [...] So the best advice I can offer the young person is to take the self out of it and realise that it's all about relationships.
>
> (Massey 2009, p.11)

In a general understanding, such connections do not negate the singular producer within those wider relationships. This operation might result in gain through the old adage that 'it's *who* you know not *what* you know', but if we widen the notion to 'those that produce the product', which is by and large different from the 'record producer', then it holds true that there are multiple producers of the artwork. Following on in the interview from Niebank, Tony Brown (Massey 2009, p.12) notes how some of those that are "more talented than me" are not producing significant work because they keep to small circles of relationships, and "in fact, they're all pissed off and they have a chip on their shoulder. I just keep widening my circle." As songwriter and producer Bruce Woolley told our students at York St John University "Collaborate. Find the people who can do what you can't." If you want to record the sea, you may need to take someone with you to carry the recorder, and find someone to plug it in, and you'll need the sea as well, perhaps a banana in case you get hungry, although you might leave that in the studio for later, it might act on you to make a different production decision to the one that the can of beer might have led to.

3 Creativity and the Production Habitus

Christopher Johnson

Overview

This chapter considers how the creativity of people involved in a music production project is affected by the music production process. It investigates how different models of collaboration and different approaches to the technology and environments involved in the production affect the aesthetic of the final produced work. In particular, it examines the links between the *personality* of the production process, and the *personality* of the final product. In attributing personality to a production process, we are concerned with the array of social, environmental, collaborative and technological factors that combine to lend a project a particular impression, or vibe, that is experienced by its collaborators whenever they are working on it, or even whenever they engage with it, however briefly between sessions. This project culture contains its own doxa, its own set of rules, codes, expectations, relationships, and methods (its habitus) that are generated by the collaborators and managed by the producer. We argue that it profoundly influences the creative contributions of the collaborators, from motivating positive attitudes to the project to nurturing creative performances in the desired aesthetic direction. We investigate how the vibe of this project culture, or the *personality* of the production process, is present in the vibe, or *personality*, of the final product.

As much as production practice centred on the Digital Audio Workstation (DAW) allows a solo creator a great degree of artistic control, there remains the creative need for collaboration on musical productions in many projects. If a producer–songwriter is to nurture their intended project themes and meaning through the production process, they must be able to communicate it effectively to their collaborators, explicitly through language or implicitly through the strength of meaning carried in their own contributions. For example, in considering the choices made by a mix engineer, Marrington discusses their responsibility to identify and align with the intentions of the artist and so bring their own creative and technical skill in delivering that artistic vision (2016, pp.199–210). Therefore, in managing the contributions of collaborators, questions of artistic, social, and commercial best practice are presented for the producer. For instance, if a musical performance is present in a recorded product then it is representing a musician's creative contribution. However, if the part being played is composed by the producer–songwriter to be performed by the musician and then edited and manipulated in the production process, the creative personality of the musician being represented is affected, and so too their motivation to take part. The social context of many bands traditionally holds that the musicians each compose their own parts, and their performance of the part is imbued with their authentic personality and contains their individual statement of meaning that contributes to the overall perception of the track. If this process is hampered by the mediation of the technology, the sonic tinkering of an engineer, or the requirements of a songwriter–producer who has written all the parts themselves and views them as essential components of the track, then what happens to the value of the musician's contribution? How do these considerations in the production process affect

DOI: 10.4324/9781351111959-4

the character of the finished product? In such a scenario, the producer, enabled by the DAW to assemble all such elements alone, must be careful to leave space for the contributions of their fellow collaborators or face them losing interest in a project that does not represent their individual artistry adequately enough, and being limited by the lack of input from other creative individuals. The production model, therefore, must be carefully constructed to consider and enhance the defined artistic persona of the band; the individual creative self of each collaborator; the artistic temperaments of the band members; the motivation of each collaborator; the expectations of the intended audience; and the meaning of each song.

In studying creativity in music production, McIntyre adopts Csikszentmihalyi's systems model of creativity (McIntyre 2009, 2012, 2019). This sees creativity happening in the interrelations of a three-component cyclical system consisting of a cultural body (the domain), a social group (the field), and an individual. The domain is a body of knowledge containing a set of rules, codes, expectations, and requirements particular to that domain, for example the domain of pop music. The individual – let's say a pop songwriter – having been steeped in this domain, has learnt its rules so well that they become second nature. They have developed a 'feel for the game' and know instinctively where the middle eight should come. This individual is then able to use their domain skills and knowledge to create something novel, like a new song, and offer it to the field, a social group who hold the knowledge of the domain and act as its gatekeepers – for example, pop fans, commentators, record labels, and music journalists. If the field accepts the individual's new work as 'a new pop song' with something fresh about it, it enters the pop domain, which is then altered by the novel addition, and in this way the domain is able to grow and evolve (Csikszentmihalyi 2013, pp.28–29). There is an interesting point about limitation here. In examining the relationship between entrenched domain knowledge, a social field, and the prospective creative individual, Kerrigan notes that "an individual's ability to make creative choices is both enabled and constrained by the structures they engage with" (2013, p.144). In our pop song example, the composition and production choices available to the artist are limited by the underlying aim of having the resulting music accepted by the pop audiences and commentators. Simultaneously, such external defining of boundaries can be necessary to enable a creative response for the producer–artist, working alone with their internalised collaborators (see Chapter 12).

However, a genre-based illustration of the systems model is overly simplistic when scrutinising collaborative creative contributions of ensemble projects, and when studying the complex social, environmental, musical, technological, and psychological factors involved in music production. McIntyre clarifies that rather than a generalised system such as the one described above we are instead dealing with a complex network of interconnected systems:

> From this perspective there is an engagement with a deeply multifactorial, non-linear, interconnected and scalable set of systems which extend vertically, up and down, system within system within system. These are also deeply and horizontally interconnected with related domains and fields.
>
> (McIntyre 2019, p.70)

He draws on the ideas of Arthur Koestler (1975 [1967]) to express these interconnected systems as *holons*, identifiable whole parts with their own unique identities that are themselves made up of subordinate other parts and that in turn simultaneously constitute part of a larger whole (Edwards N.D.). This perspective of proximate nested systems creating a complex network of inter-related domains and fields is more useful to us in examining the collaborative creative process of music production. It views creativity not as the result of a single system but of many inter-related and intersecting ones. It presents the discipline-specific skills, knowledge, instincts and aesthetics of each collaborator and their influences as distinct wholes that are brought together for the purposes of the project, and just as a mix engineer applies aesthetic judgement to the manner in which they bring together individual audio tracks, so a producer must apply considered artistic judgement to the manner in which they bring together the collaborators and the creative systems, or holons, that come with them.

We are going to investigate how this deliberate assembling of these constituent whole parts and their accompanying systems is performed by the producer as they select the personnel and organise the conditions for their production.

Bourdieu and Habitus

To examine the relationships between the skills, habits and dispositions of each collaborator, and the culture that is set up by the producer for a project, we can use Bourdieu's conceptualisation of habitus (1977), briefly summarised by Costa and Murphy:

> Habitus is socialised subjectivity that agents embody both individually and collectively, through the interrelationships they establish in the social spaces to which they belong. Habitus encapsulates social action through dispositions and can be broadly explained as the evolving process through which individuals act, think, perceive and approach the world and their role in it. Habitus thus denotes a way of being.
>
> (Costa and Murphy 2015, p.7)

Bourdieu proposes habitus as a model for understanding society, a construction of how society in all its structuring rules and prevailing habits operates between people. It refers to "the set of dispositions and meanings that people gain through socialisation" (Gaventa 2003, p.8). It is in the relationship between social structures and the tendencies of behaviour and thought that exist within them:

> [Habitus…] expresses first the result of an organising action, with a meaning close to that of words such as structure; it also designates a way of being, a habitual state (especially of the body) and, in particular, a predisposition, tendency, propensity, or inclination.
>
> (Bourdieu 1977, p.214)

As such we can understand habitus in terms of the relationships between a person, the rules of the social environment they inhabit, and others operating in the same group. The shared actions, conceptions and world view that emerge from the rules of the social structure have an effect on emerging dispositions of those in the group (Smith 2020). As a leading interpreter of Bourdieu's work, Wacquant describes the concept as a mediating construct, that is "the ways in which the socio-symbolic structures of society become deposited inside persons in the form of lasting *dispositions,* or trained capacities and patterned propensities to think, feel and act in determinate ways" (Wacquant 2016, p.65).

Our music producer, in drawing together the collaborators and organising the spaces and resources of a project is in effect arranging a micro-socio environment where the effects of habitus can be considered to shape the aesthetic of the project. Although Wacquant makes it clear that habitus "is not a self-sufficient mechanism for the generation of action" (2016, p.64), the producer can have an awareness of the existing effects of habitus in their collaborators and make their production decisions accordingly and creatively.

Wacquant identifies and examines two types of habitus. Primary habitus is to do with early childhood experiences and the social habits, knowledge and understanding that is acquired through immersion in familial environments with no deliberate effort of learning. A secondary habitus, on the other hand, is "any system of transposable schemata that becomes grafted subsequently, through specialised pedagogical labor that is typically shortened in duration, accelerated in pace, and explicit in organisation" (Wacquant 2014, p.7), such as the habitus acquired through studying and practicing music technology. This secondary habitus will be crucial to the music producer as they selectively assemble their actors.

For Bourdieu, a core concept of habitus is the capital that exists in social resources, be it material, cultural, social, or symbolic capital (Navarro 2006). A resource that has a social power value or status, functions as capital. If it has a cultural or symbolic significance that affects the social structures and the way people operate around it, it has capital. For example, Abbey Road Studios has a cultural capital that affects the behaviours of musicians that record there. A well-connected music manager could have social capital due to the network of influential people they have access to. Social and cultural capital are significant in a competitive society as non-economic forms of power, and for Bourdieu the struggle for power is central to all social arrangements (Navarro 2006). Therefore, a music producer with an awareness of the capital inherent in their collaborators, and in the objects, institutions, and physical environments brought into a project, can have an influence in shaping behaviours and creative responses. This includes the channels selected for communicating messages within the project, as some messages benefit from the weight they acquire when communicated on an emailed document with a well-formatted header due to the inherent symbolic capital. Other messages will be better communicated with the less formal, lighter touch of a personal text message, where the deliberate absence of certain types of capital communicates its own meaning that extends the words in the message.

Capital and how it is awarded value and distributed in a social group is an important structuring element in defining one grouping as distinct from another. This is another part of Bourdieu's theory, the idea of 'fields.' These are the social and institutional arenas in which people compete for capital, and where they exhibit the dispositions of their habitus (Gaventa 2003, p.9). For example, a successful music producer may have cultural and social capital in the field of the music industry, earned through a catalogue of respected work and the network of contacts they have built up throughout their career. They would hold a power that other members of the field, other people working in music production, would recognise and respond to. The same person would be unlikely to hold the same capital when operating in a neighbouring cultural field, say that of television production, and so the limits of the capital, or power distribution, can define the limits of the field.

Related to habitus is Bourdieu's understanding of 'doxa', which refers to taken-for-granted norms and beliefs, or 'common sense'. As a set of unchallenged assumptions, doxa is both a source and a manifestation of power (Gaventa 2003). It happens when the limits of power are constantly encountered through life and so become social divisions, ingrained patterns of thought that inscribe as a social order, "a 'sense of one's place' which leads one to exclude oneself from the goods, persons, places and so forth from which one is excluded" (Bourdieu 2010, p.473). For Bourdieu, doxa is "an adherence to relations of order which, because they structure inseparably both the real world and the thought world, are accepted as self-evident". If habitus is a construction of how society operates and the way people operate within it, doxa can help us understand why.

In his research into creativity and popular music McIntyre (2009, 2019) uses 'habitus' in relation to the specialist internalised habits and knowledge an individual has acquired, much like Wacquant's secondary habitus, and the opportunities to apply them in a 'field of works'. He presents it as a similar concept to the individual being immersed in the body of knowledge of a domain in Csikszentmihalyi's systems model. However, the notion of a short-lived secondary habitus is also useful to us at a micro level in conceptualising the production culture that emerges for a single specific project, and that the producer attempts to shape. The overall creative social space generated in which participants become aware of 'ways of doing things' that pervade the project and in a large sense, define it. In using habitus and Bourdieu's related concepts of capital, fields, and doxa as tools to explore the production culture of a project we can scrutinise the multiple agents, systems and holons at play, "taking into consideration its complexity as a container of practices imbued in the objective and subjective contexts of the phenomenon under study" (Costa and Murphy 2015, p.8).

The Production Habitus

I propose the term Production Habitus to refer to the composite creation of the producer when they 'write' the logistical production process; assemble the team with their internalised domains, doxa and inherent capital; stimulate creative motivation in the participants; form and manage the social environment and micro culture surrounding a project; and select, design or manipulate the physical production spaces. It involves a conceptual bubble that surrounds a project; notions and ideas that permeate it; capital and habitus that are expressed within it; emotions that the participants feel whenever they interact with it. It affects their motivations, cognitive processes, behaviours, and contributions in a way that the essence of the Production Habitus becomes embedded in the final product. It is useful here to consider John-Steiner's integrative type of collaboration (2000, p.213), which describes an intense, short-lived form of collaboration based on commitment to a shared vision. In many fields the social rules and cognitive processes attached to these collaborations emerge organically. The notion of a Production Habitus in music production describes the deliberate shaping of these factors to create an aesthetic quality to the collaborative project that becomes embedded in the resulting product. For Burgess' (2013) Enablative or Consultative producer types who are characterised by a hands-off role when it comes to the composition, performance or engineering of a project, and who are more concerned with "sourcing the talent and material and creating conditions in which a successful recording could take place" (Burgess 2013, p.15), the creation and maintenance of the Production Habitus could be their primary, and crucial creative contribution.

However, if the producer wishes to create the best conditions for creative contributions within the Production Habitus of a project, they will need to consider wider theories of creativity beyond the individuals involved, their nested creative systems, and the effects of capital and doxa.

Considering Conditions for Creative Contributions

In researching the social psychology of creativity, Amabile has suggested other key influences of creativity focusing on external events from the social environment, personality traits, generic creative skills that can be taught, and factors that affect an individual's motivation in approaching an activity. She proposed the Intrinsic Motivation Principle of Creativity (1983, 1996, 2013), which states that "the intrinsically motivated state is conducive to creativity whereas the extrinsically motivated state is detrimental" (Amabile and Pillemer 2012, p.7). Across multiple studies it was observed that participants were more creative when motivated by the act of doing the task itself. They would derive value in finding it satisfying, stimulating, fun, or interesting, and this would enhance creativity. Lower levels of creativity were found when participants were more extrinsically motivated, for example by payment, the promise of reward, the prospect of evaluation, or the awareness of surveillance. These findings suggest a detrimental effect of certain types of capital on creativity. If the motivation is based on material capital in the form of monetary reward, or the cultural or symbolic capital that may be transferred or awarded by another in the form of praise or evaluation, then creativity suffers. This implies a complex relationship between capital and creativity. On one hand, it suggests that people are more creative when given freedom from the daily struggles of competing for capital in Bourdieu's fields, that they can express themselves better outside the doxa of their usual habitus. On the other hand, the acquisition of cultural capital by the validation of one's works by respected others is a core motivation of the musical artist presenting work to the field for consideration, such as in the systems model.

This relationship between intrinsic motivation and creativity has been further investigated by Liu, Chen, and Yao (2011, in Amabile and Pillemer 2012) who found harmonious passion to be a more effective motivational mechanism. Harmonious passion, as proposed by Vallerand *et al.* (2003), refers to the internalisation of an activity that a person enjoys doing and feels is important to their life to the point where it becomes part of their core identity, fulfilling a basic human need for autonomy and freedom of choice. An example would be a person choosing to spend their time regularly playing the guitar in the pursuit of musical excellence, and so having 'guitar player' become part of their core self. Liu, Chen, and Yao (2011) found that harmonious passion, as a stable form of intrinsic motivation, can be a strong predictor of creativity. Therefore, in arranging the logistical structures and personnel in a music production scenario, the producer seeking creative contributions from the project participants should prioritise 1: engaging them in activities in which each participant is personally intrinsically motivated, and 2: making space for their harmonious passions to be practised.

Amabile integrated the principle of intrinsic motivation into the Componential Theory of Creativity (1983, 1996, 2013). This model describes four components that are required for a person to produce creative work. Three of these are psychological components of the individual. The first is the domain specific skills and expertise of the individual, such as their technical aptitude in the particular domain. The second refers to their creativity-relevant processes, which include personality traits and cognitive approaches to problem-solving, flexible thinking styles, and open-mindedness. Some of these skills and related techniques can be taught and developed, such as mind mapping, reverse thinking, and brainstorming (Smith 1998). The third psychological component is intrinsic task motivation, as discussed above. The fourth component, external from the psychological components, is the social environment, which we can consider through our examination of habitus. Amabile (1983) cites accounts of how seemingly mundane events or details in the immediate social environment have had detrimental effects on creative work: "For example, in a letter to a friend, Tchaikovsky (1906) described the devasting effect that simple interruptions could have on his work" (Amabile 1983, p.357). Similarly, Einstein reports becoming de-motivated by the structures of formal education:

> the fact that one had to cram all this stuff into one's mind for the examinations, whether one liked it or not. This coercion had such a deterring effect upon me that, after I had passed the final examination, I found the consideration of any scientific problem distasteful for an entire year.
>
> (Einstein 1949, p.17, cited in Amabile 1983, p.357)

Research into the social environment has focused largely on the social structures of organisations and their effect on the motivation of employees. This social environment includes those extrinsic motivators that have been shown to have a detrimental effect on creativity, such as evaluation, reward, praise, and surveillance. It can also include factors that stimulate intrinsic motivation and lead to more creativity, such as positive challenges, diverse collaborations, and open work structures that promote freedom in carrying out tasks (Amabile and Pillemer 2012).

The Componential Theory of Creativity provides a useful framework for examining how a music producer makes creative choices about the team they assemble and the social environment they create around a project through the messages implied by their communications, behaviours, their own creative contributions, and their approaches to leadership of the project. In a general summing up of the theory Amabile suggests that "creativity should be highest when an intrinsically motivated person with high domain expertise and high skill in creative thinking works in an environment high in support for creativity" (2013, p.136).

However, the theory does not address the influence of other environmental factors that are important in music production, such as the physical spaces that form a major part of the social, psychological, and sonic tone of a project. Partly in response to the Componential Model's limitations in regard to the physical environment, McCoy and Evans (2002) investigated the influence of specific interior design elements on creativity. The study identified five characteristics of the physical environment that predicted greater creativity: "(a) complexity of visual detail, (b) view of natural environment, (c) use of natural materials, (d) with fewer cool colors used, and (e) less use of manufactured or composite surface materials" (McCoy and Evans 2002, p.409). These findings are interesting when considering the design aesthetic and lighting tendencies of recording studios, especially where the choice of studio, or indeed the design of a temporary production space, is a critical creative contribution of the music producer.

We will also need to focus more on the aural environment, particularly as experienced by musicians as they perform. For example, Williams (2012, pp.113–127) notes how the aural space of a headphone mix has a particular effect on how intimately musicians respond to each other during performances, and how the studio environment can set up a sort of hierarchy of monitor mixes with individual musicians' mixes becoming subordinate to the engineer's control room mix. In the habitus of this common recording studio situation the engineer has an unspoken cultural capital that gives them power in their studio field, and some of this capital is transferred to the control mix. The musicians have less capital whilst in this specific studio field, and so do their headphone mixes. Such social power lines can have an effect on the performer and therefore the performance, and it's not uncommon for trust issues to develop when a musician asks for a change to their mix, the engineer obliges and the musician hears no change, or worse an exaggerated change that they interpret as frustration on the engineer's part. Developments in studio connectivity have addressed this situation, with technology such as Roland's M-48 Live Personal Mixer system allowing each musician to control their own mix directly, disrupting the arrangement of capital in the studio during the tracking phase of a session. Affording some control to the musicians is giving them more space for their intrinsic motivation to activate. They can quite privately, even self-indulgently, construct their own aural world relatively free from the dominance of the engineer, spending more time in their mental creative zone by crafting their own experience. This will affect their performance.

Similarly, the way the studio space is set up to allow sightlines, or not, between participants during takes will have an influence. Musicians will perform together differently if they can see each other, for example a bassist and drummer rhythm section will be able to predict and follow accents and nuances in each other's performance if they can see the heft and swing of each other's bodies as they play. A vocalist may emote more expressively if they are aware of a small supportive control room audience, or conversely may need to feel completely unobserved in order to twist themselves in the right shapes and movements to produce the desired emotional performance. Such elements need to be considered for their effect on creativity and musical performance.

Moreover, small but potentially significant elements of the social environment can fall beneath the Componential Model of Creativity's level of scrutiny, such as Tchaikovsky's unwelcome interruptions, or the effect that having a large bowl of fresh fruit in the control room of a studio has on the musicians booked for the session. Also, by focusing on the cognitive processes of the individual, the theory does not consider their physiological state, which is something the producer can have an effect on. Working hours, accommodation, meals, break times, and supplies of refreshments are all within the control of the producer and will affect participants' energy and stress levels, as well as their general health, which is likely to affect creativity and performance.

In an earlier analysis of the study of creativity Rhodes (1961) observed an overemphasis on the creative individual and suggested an alternative multi-dimensional approach. The four Ps framework categorises creativity and directs analysis toward the *person*, *process*, *press*, and *product* involved in a creative endeavour and how creativity occurs at the confluence of these aspects (Garces *et al.* 2016). The *person* category includes the personality, attitudes, intellect, and skills of the individual. The *process* is about the stages involved in creativity, including the thinking process and the progression of an idea as it develops. The *product* is the idea when it is realised in a tangible form, although Rhodes suggests a hierarchy whereby the idea has greater creative capital than the product, pointing to legal structures around patents and copyright to demonstrate. The *press*, or environment, relates to the environment where the creativity is situated. Rhodes points out that inventions don't occur in a vacuum, but rather in response to a social need and on the back of existing technologies and components that the inventor draws together. The four Ps framework has become a main reference in the study of creativity. Jordanous argues that it enables an "inclusive and encompassing approach to the study of creativity, accommodating multiple relevant perspectives" (2016, p.165). In studying the creativity of music production and the aims of a music producer to elicit creative contributions from their collaborators, the four Ps framework provides a broader perspective than the Componential Theory in that is includes examination of the *product*, defined as the tangible embodiment of the creativity. We could also include the physiological state of the individual in its broader definition of the creative *person*. Studies of the *press*, or environment, that have used the four Ps framework have focused on the cultural, rather than the physical environment (Garces *et al.* 2016; Jordanous 2016). This suggests the concept encompasses the wider cultural domain of the systems model, as well as the more immediate social environment of the Componential Model. In Jordanous' interpretation (2016) the press/environment is bidirectional, involving the influence the social environment has on the creator whilst in the process of creating, as well as the influence the social environment has on the creator after the product has been publicised and feedback has been given. This concept could be articulated as a cycle as the creator makes things, submits them to the culture, becomes affected by the feedback, and then makes the next thing and so on. A cyclic framework is useful in considering the persona of a music producer working within a cultural system and responding to the public reaction to their works through creative input into subsequent works. We will explore this in Chapter 11.

Authoring a Process

I suggest that in bringing together the components for a project the producer dynamically authors a complex production culture and micro socio environment. This includes the selection of collaborators, spaces, domains, production methods and technologies, but consideration is also given to *how* these components are assembled. For instance, the communication methods and messages contained therein; the careful arrangement of capital associated with each component; and the focused manipulation of production goals, conditions, and environments all form part of the production culture that is composed by the producer. To state this notion considering Barthes' ideas (1977), I propose that the production process itself is a text that is 'written' by the producer drawing from a vast repository of existing codes and signs, which is then 'read' by each collaborator. The codes and signs may range from norms and beliefs of primary habitus – general systems understood by the layperson, such as language, topical themes, and logistical arrangements to those of numerous secondary habitus – the more technical or specialist domain matters able to be understood by collaborators due to their previous immersion in music and production, such as production methodologies, musical terminology, and specific acoustical or equipment details. The manner in which the producer 'writes' and controls the production culture and its processes creates meaning in the mind of the collaborator: a feeling, understanding, or 'vibe' of the project, which affects the collaborator's professional and emotional connection to it. This connection in turn affects their intrinsic motivation, behaviour and creative performance on the project and it is therefore of profound importance. Each collaborator represents a unique multiplicity of intersecting fields and domains that they have been steeped in, and that play an important part in shaping their aesthetic and their creative persona, and in influencing their contributions to a project. The producer's handling of the project's culture will involve discussions, activities, and arrangements that selectively foreground and background these internal domains, guiding the overall creative direction of the project, even when the producer is working alone and is only dealing with their own internal brew of networked cultural systems (see Chapter 12). This project culture that is authored by the producer is the Production Habitus.

To state this in terms of the four Ps framework, the producer makes creative decisions, drawing together and influencing various components to design a project for an aesthetic goal. They recruit the *person*(s) with the desired domain skills, creative thinking styles and attitudes, and stimulate in them creative behaviours and intrinsic motivation. They arrange the *processes* and stages of progression, keeping clear communications and logistics appropriate to the project's aesthetic aims through the development of the production. They arrange the *press*, social environment, or place, in their selection, arrangement or design of the physical studio spaces, and in their management of the immediate social environment surrounding the production. The results of these choices are crystallised in the *product*, whose effect on the wider social environment, or domain, is perceived as the end goal during the production, and whose ultimate existence represents the embodiment of the creative efforts of the participants, stimulated, guided and combined by the producer.

Foucault (2011 [1969]) proposed the notion of 'transdiscursive' authors to explore the concept of individuals who author "a theory, tradition, or discipline in which other books and authors will in turn find a place" (2011, p.534). Without claiming quite such an impact in the 'writing' of music production processes and micro cultures, the notion is still useful in proposing that the considered assembling and manipulation of elements in a highly individualised music production scenario can be understood as an authoring process that produces "the possibilities and the rules for the formation of other texts" (2011, p.534). That is, predictable patterns could emerge in the activities of the producer; repeatable models of effective practice in a Production Habitus that can be modified for subsequent use in other projects.

It is worth noting at this point that the producer's 'writing' of the Production Habitus is continual and dynamic throughout the production, responding to changing situations and the aesthetic progression of the work accordingly. Much of the core project design will take place in the early stages so that the overall logistics and aesthetic direction are established, but there will inevitably be room left for adaptation and variation to allow collaborative creativity to flourish. We want to highlight that this dynamic writing process is more than just the logistical tasks of booking resources and personnel. We are more interested in the creative act of bringing such things together for a particular artistic goal; what lies behind each of the producer's decisions and *how* each task is performed in a deliberate manner to stimulate meaning in the mind of each 'reader,' or collaborator.

The producer's 'writing' of the Production Habitus is therefore critical in creating the aesthetic of the product. Their assembling and shaping of the network of creative systems, or holons, will be influenced by their understanding of each interlocking part and its structures of cultural and social capital. This implies that a producer is advantaged if they have a broad understanding of each subset domain or secondary habitus, and how they interface and interlock, such as songwriting, drumming, sound engineering, studio management, coaching, programming synths, singing, scoring for strings, tuning up guitar amps, and so on. And indeed, an open understanding of what they don't understand about each constituent field and holon will be equally important in composing a creative production culture that maximises collaborative creativity whilst directing the project toward its aesthetic aims. Viewed through the four Ps framework (Rhodes 1961), which examines creativity in the confluence of person, process, press/environment, and product, the deliberate crafting of the process and press, plus the careful assemblage of the persons will have a clear effect on the product.

Figure 3.1 illustrates some of the core constructs that make up the Production Habitus. They are all elements that are considered, selected and dynamically assembled by the producer in a deliberate manner to create the conditions for a specific aesthetic result. They create the cloud of understanding around a project, affecting the psychological, creative and emotional spaces within each collaborator which will emerge in their creative contributions and ultimately be embedded in the final product. The concept synthesises ideas presented in established theories of creativity. The '*person*' from Rhodes' four Ps theory (1961) is extended to '*persons*' and includes consideration of the holonic arrangement of nested domains internalised within each collaborator (McIntyre 2019; Csikszentmihalyi 2013). The *persons* element also considers each collaborator's creative thinking skills, as suggested in the Componential Theory of Creativity (Amabile 1983), as well as their physiological, psychological and emotional states. Rhodes' '*process*' (1961) is incorporated into the Production Model element, in how the production tasks are allocated and sequenced, and in how the collaborator's contributions are arranged, received and managed. In using the concept of a Production Habitus, the '*product*' from the four Ps framework has an equivalent in the Project Definition element. Once established this will guide all the other elements. More than the physical audio product alone, the Project Definition includes the wider aims of the project, such as social or commercial intentions, as these join with the visualised product in affecting the creative contributions of the collaborators as they work toward common goals. Rhodes' '*press*' is a similar concept to Amabile's 'Social-Environment' component (1983), and in creating the Production Habitus these are split into two contributing elements. The Social Micro-Environment concerns the conventions, rules, interactions, and communications that emerge at a specific project level, rather than in the wider social domain. Including the Physical Environment element allows the producer to consider the effects of the studio layout and design, as well as the arrangement of things in the space and how these things can be manipulated to affect creativity for a specific aesthetic audio result. Amabile's Intrinsic Motivation Principle (1983) is considered in both the Motivations and the Ownership elements. The motivations of the collaborators can be considered from

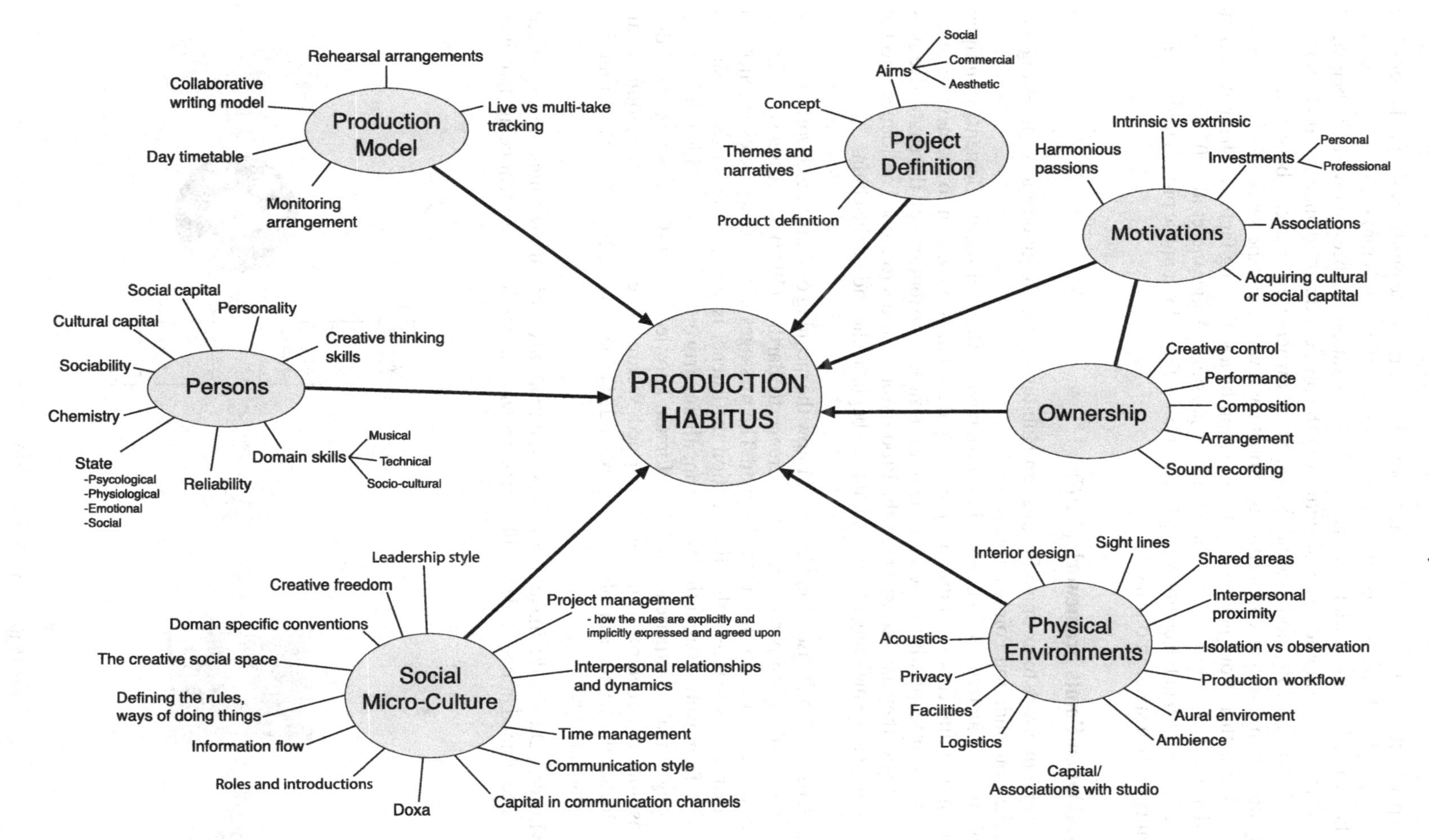

Figure 3.1 Mapping the Production Habitus

several perspectives, and in creating the Production Habitus the producer should be sensitive to the personal and professional investments each collaborator is making to the project and how they will affect motivation. Whilst intrinsic motivation has been shown to be beneficial for creativity, the legal and social structures in the music industry demand that Ownership be considered as a core motivational factor. Some ownership issues can be seen as extrinsically motivating in their connection to monetary reward, but the sense of ownership an artist has over their work goes beyond this. It can be a validating factor in the make-up of their identity, a core part of their harmonious passions (Vallerand et al 2003). It is a separate consideration in the creation of a Production Habitus for a music project and therefore warrants its own space in our model.

Consolidation Points Framework (CPF)

A helpful framework to examine Production Habitus, or even to guide efforts in setting one up, is the Consolidation Points Framework.

This focuses analysis on moments in a production where disparate elements (multiple) are assembled and combined into a whole (singular) that is passed on to the next stage. In doing so, our attention is focused on the non-musical contributions of the enablative producer and how they have constructed the Production Habitus to embed aesthetics into the product and create meaning for the listener. To begin at the end of the chain:

Listener: We will discuss in Chapter 11 how the meaning of a produced work is created in the mind of the perceiver, with reference to Barthes (1977) and Eisenstein (2010). For a music product, this is the listener. The messages in the audio are experienced in tandem with whatever other information the listener is aware of about the music and artist, from word of mouth, journalism, album artwork, and so on. This is all filtered through their life experience, habitus, domain knowledge, world view, and general disposition, and a composite meaning is formed. We are interested in how the producer can influence the elements involved in this expanded experience of the listener encountering the final product.

Master: During the mastering process the final mixes of all the songs are sequenced and combined into a single tangible product with the intention of them being perceived

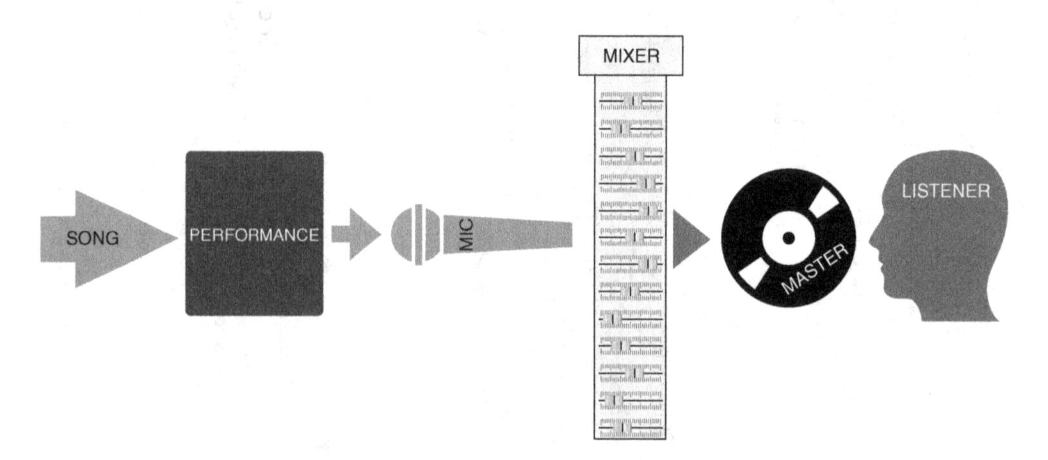

Figure 3.2 Consolidation Points Framework (CPF)

in the singular: the 'album'. The mastering engineer is only included in the Production Habitus at a late stage and will need a condensed briefing from the producer to guide how they shape the audio processing and track sequencing to create the desired audio experience of the album.

Mixer: The most overt representation of combining multiple elements into one in music production practice. The mixer (the person and the console) takes the audio from each channel and balances it into a single audio rendering of the song that we can call in the singular, the 'mix.' However, what is being blended is not just audio. It includes personalities, moods, relationships, and meanings that have been expressed in the performances and programming. The mix process includes the deliberate selection, foregrounding, or blending of all these conceptual elements. Mixing is such a complex process with infinite creative variables throughout that the mix engineer needs guidelines and constraints to understand the project and properly achieve their mix goals. Many of these general guidelines come from the mixer themselves, from their personal aesthetic, itself the composite of multiple internalised domains from mix practice, audio engineering, the wider domain of contemporary music, and so on. However, the more specific constraints will come from the mix engineer's understanding of the project before it gets to them and its aims once they pass on the mixes. These things are communicated within the Production Habitus and are usually in the control of the producer. At the mix stage, the production team is often reduced to just mix engineer and producer, so it is in the role of the producer to maintain the Production Habitus and induct the mix engineer into it.

Mic: The mic captures and combines many elements into a single signal that is routed to the mix console and/or recording device. By examining these elements and assessing the control the enablative producer has over them, we can better understand the construction of the social, technical, and physical environments that form part of the Production Habitus. To illustrate, we can consider a single kick drum mic and what it captures, starting with the drum itself. Choices are made over the drum's dimensions, materials, manufacturer, beater, tuning, dampening, skins, and overall condition that are within the producer's control as they source the instruments for the session beforehand, and then collaborate with the engineer and the drummer when setting up on the day. Then there is the physical acoustic environment the drum is in, how the room is arranged for performance, the position of the drum in the room, and the position of the mic on the drum, which will all have been agreed in collaboration with the engineer within the Production Habitus. The mic will also capture bleed from other instruments that are recorded in the same room at the same time as the drums, which is a result of decisions made in the production design and process. The producer will have arranged for instruments to be recorded separately or together; for some musicians to be able to perform in the same room and others to be isolated. The drum sound captured by the mic is a consolidation of these factors, with various people's skills, aesthetics, and judgements at play. They are collaborating within the Production Habitus and are managed towards a specific goal by the producer. The mic represents a physical node in the Production Habitus where all these carefully considered factors are rendered into a single audio signal.

Performance: This is perhaps the Consolidation Points most susceptible to influence from the Production Habitus. It combines the skill and artistic personality of the performer with their understanding of the project and their general state in the session. They will be affected by emotional, psychological, social, and biological factors, and these will in turn affect the performance. It is within the control of the producer to design a Production Habitus that is aware of many of these factors: removing potential stressors; providing refreshments; ensuring adequate breaks; making space for people when they need it; controlling the ambience and social vibe; recruiting suitable personalities for the project; being sensitive to social capital and power relations; keeping everybody informed and generally making sure everybody is looked after. Other logistical matters of production design will find their way into the performance, such as how many people are present to witness the performance, the time of day the performance happens, and how long is allocated to get the performance right. Monitoring is another important factor. For instance, the decision for a performer to monitor static elements such as a click track and pre-recorded guide tracks, or to record more things simultaneously and monitor the live performances of the other musicians, will have a major effect. Similarly, having everybody share a single mix can produce more sympathetic performances than everybody having separate mixes, which can alter their focus and lead to more self-conscious performances within an ensemble. On the other hand, we explored earlier how giving a musician control over their monitoring can have a liberating effect from a studio hierarchy that can facilitate creative performances. Altering the balance and vocal processing in a singer's headphone mix can fundamentally change their performance. Some artists benefit from time and physical space to 'get into character' leading up to their performance, where they can do the creative and psychological work required to focus on the meanings and persona they wish to express. Many respond to subdued ambient lighting in the studio or interior design features that mark the studio as a safe creative space. All of these are elements of the production process, part of the Production Habitus that can be considered and manipulated for the project aims. They are consolidated within the performer and emerge as a single event – the performance.

Song: Fundamentally, the song can be considered to combine lyrics, harmony and melody into a single format that communicates meaning. On some projects, the songs will be the starting point and the themes of the Production Habitus will emerge from them. On other projects, the themes of the Production Habitus will inform the songwriting or the selection of songs. More than just fitting an album concept, the songs will be selected or written with the logistical limitations and opportunities of the project in mind. The persons and organisations attached to the project could also affect song selection by way of their symbolic capital. In our CPF framework the 'song' also includes the arrangement which combines the separate instrumental parts into a single musical piece. The effect of the Production Habitus on the arrangement varies greatly on the project. In democratic band projects where the parts are worked out collaboratively between the members, a form of Production Habitus will exist before the project reaches production. Social dynamics and power relations; favoured rehearsal spaces; time commitments; live performances; arrangement habits; and access to technology will contribute to a definable 'way of doing things' that drives how the band develops material. A producer coming in to collaborate with such a band can use the existing habitus as a starting point for the Production Habitus, or even dispense with it deliberately to refresh the band's creativity. In short, the raw musical materials of melody, rhythm and harmony are articulated in the arrangement and combined with lyrics into a single entity that we encounter as the song.

The Consolidation Points Framework is intended as a tool to focus analysis when practically dealing with the multi-layered constructs and concepts of the Production Habitus. It should facilitate consideration of all the key points of a production in a more structured and accessible manner.

4 The Production Habitus of *Smoke Rainbows – Music Minds Matter* (Abbey Road Case Study No.1)

Christopher Johnson

The previous chapter proposed Production Habitus as a way of understanding the non-musical contributions of music producers. We examined the considered assembling of people, environments, and social conditions to create a culture that would best suit a project's goals and produce the desired outcome. To investigate this concept practically we devised a collaborative studio-based music production project as a case study that this chapter will examine. The coproducers' roles would not include songwriting, arrangements, or performance so we could instead focus on the creation of the Production Habitus and observe its effect in action. Much of the work to establish the Production Habitus happened in the early planning stages of the project and by examining the considerations of the coproducers and the choices they made we can analyse the conscious design of the Production Habitus, how it was shaped and maintained, and how its essence is rendered in the resulting musical product, the album *Smoke Rainbows – Music Minds Matter* (2019).

Creating the Production Habitus: Project Initiation

Certain elements for the environment of the production were in place before the project was devised. That is, every year the York St John University Music Production students take part in a week of professional projects at Abbey Road Studios as part of their developing production experience. Studio 2 in Abbey Road was already booked for this, which is of course world famous as the studio that produced nearly all The Beatles' output as well as landmark recordings by Pink Floyd, Oasis, Radiohead, and Kate Bush (Abbey Road 2020). I was able to secure four days for this booking in 2018, which meant the dates in February, the prestigious studio location, and the presence of students were fixed elements in the planning that I would be able to design a project around.

A number of key drivers seemed immediately important that would affect the kind of project I could devise to maximise the exciting opportunity of having Abbey Road Studio 2 for four full days. First, the project needed to produce a real commercial product such as an album that would be physically released. Defining the product commercially early in the process would set a target for everybody and mitigate against the possibility of the project being perceived as less professional due to the involvement of students. Second, it should provide opportunities for talented artists who wouldn't usually get the chance to record at Abbey Road. A result of the mechanics of commerce in the music industry is that many highly talented individuals are never commercially successful enough to get the funding or support to perform in such a prestigious venue, and so this project could enable them to demonstrate their talents at a high level. Third, it should have a moral, human angle and aim to somehow contribute something positive in the world. This is partly from a personal sense of social responsibility when given opportunities like this, but it would also play a part in defining a common motivation amongst all the participants and a solid theme that could

DOI: 10.4324/9781351111959-5

focus their contributions and help define the product and shape its release strategy. It would be an element to bring everybody together emotionally and conceptually, something to give powerful cohesion to a collaboration of many people who do not yet know each other. Finally, it needed to provide a valuable hands-on experience for the students.

The fact that these thought processes were happening even before the project was devised shows that the intrinsic motivation of the participants was already being considered as the Production Habitus was conceived. The social environment of the project was deliberately being formed with positivity, camaraderie, and social justice as guiding principles that would influence the behaviours, motivations, and creative contributions from that point on. In our map of the Production Habitus (Figure 3.1, Chapter 3) it is the Project Definition component that is being created: the commercial, social, and aesthetic goals of the project, along with the vision for the final product.

The initial idea drawn from the guiding principles was to partner with Karousel Music, a not-for-profit collective of artists, music management, and publishing, to produce a commercial album that would give 12 talented emerging artists on their roster the opportunity to record in Abbey Road, Studio 2. They would rehearse with a house band in preproduction rehearsals and perform one song each live in the studio for a 'Live at Abbey Road' style album (the terms "Live at Abbey Road" and "Live From Abbey Road" are protected intellectual property and therefore we couldn't use them in the actual title of the album. We are only permitted to use the name "Abbey Road" as a factual reference to where the recordings took place). The house band format would give the producers some control over the musical arrangements and ensure a sonically coherent product. It would also go some way in addressing the logistical challenge of recording an album with 12 different artists in just four days, as the band and backline would remain relatively static for each artist. The album would be for the benefit of the Music Minds Matter campaign (see, for example, Marshall 2018), which is a mental health awareness and support service for musicians run by the charity Help Musicians UK.

This proposal was pitched to Chris Sheehan, the director of Karousel Music, who agreed to join the project as coproducer. The idea was developed in subsequent discussions between the coproducers before approaching any artists. It was understood that we had to have a fully developed idea of the project before pitching it to any artists. We wanted to be able to describe the finished album, outline how the sessions would run, detail the commitment required of everybody, and give an impression of what it would feel like to be involved. This meant some extensive aesthetic and logistical planning but also close consideration of the position of the artists, processes which we can understand as designing the Production Habitus – the practical, creative and social bubble that the project would generate and thus inhabit. The first communication with potential artists would be crucial in shaping their emotional perception of the project, as the coproducers extended the Production Habitus beyond their initial pre-production inner circle. Understanding this importance meant that the first contact with artists was seen as a major milestone in the production, with the planning ramping up towards it.

Developing the Production Habitus: Pre-Production

The initial house band format was dropped early on, partly due to logistical difficulties in arranging adequate pre-production sessions over the Christmas period, but more importantly to widen the opportunity to artists with their own bands and to allow them greater control and ownership over their contributions. This transfer of creative control from producer to artist is again linked to notions of the artists' intrinsic motivation, and in retrospect it would have been difficult to ask independent artists to agree to alternative arrangements of their songs and an unfamiliar band without a more collaborative and time-intensive pre-production process. However, the responsibility for being properly rehearsed would now be with each artist rather than the producers, which would mean careful construction of the Production Habitus so that it just became understood as a 'rule' within the project's socio-environment that you arrive at the studio well prepared and very well-rehearsed. This implied 'rule' can be linked to Bourdieusian notions of power and the inherent cultural capital involved in a music project involving multiple artists. Although the sessions would be carefully set up as nurturing rather than competitive environments, the artists' efforts to perform their part well in the project and to be judged favourably by the collaborators and eventual audience, could be interpreted through Bourdieu as a covert attempt to gain cultural capital (Navarro 2006). Demonstrating artistic skill under pressure and in the culturally significant environment of Abbey Road could acquire for an artist a measure of power in their particular field within music. However, this perspective needs to be balanced with an artist's intrinsic desire to simply play well, as well as the effect of their morals on their actions in a project with charitable aims. Bourdieu has been criticised for downplaying the role of personal ethics and morals in social judgements (Lamont 1992; Sayer 2005; Ignatow 2009).

The Production Habitus would now have more of a role in addressing the issue of how to combine so many diverse artists into an album with a coherent sonic personality, given that the producers' pre-production input into the musical compositions and arrangements would be entirely hands-off. The alternative logistical arrangements to make the quick turnaround between acts possible was to arrange a static backline, instead of a house band, that all the artists would share. This could be pitched to them by likening the studio floor to a festival stage, where bands share the drum kit and amps. The language here taps into the doxa of festival performance and again forms part of the Production Habitus, the artists' domain awareness evoking notions of camaraderie and a team spirit, the excitement of live performance, and hectic but stage-managed changeovers, as well as prompting festival etiquette like starting and finishing your performance within your allocated time slot. The selection, cost, and high quality of the backline would be taken care of by the producers, but with the technical details and an element of choice around guitar amps and keyboards communicated to the artists well in advance so they could feel confident in the project and be keen to take part.

I had initially thought that the opportunity to record for free in Abbey Road would be enough motivation for many artists, but in considering their position more closely we realised that we were asking quite a lot of them. They would need to make themselves and their musicians available midweek, which could cost in days off work, childcare, travel, and musician fees. We were also asking them to spend enough preparation time with their bands to be able to perform instantly brilliantly in front of seasoned Abbey Road engineers and a group of unfamiliar students pretty much as soon as they hit the studio floor, which is at best intimidating and at worst a significant barrier to creativity that we would have to manage. And we were not paying them for the sessions. This dynamic made us think more generously about the offer we could make to the artists, and this in turn affected the timetabling we were drawing up for each day. Twelve artists across four days meant we would have to produce three artists per day. If we worked from 10:00 until 19:00 we could give each artist 3 hours whilst still allowing time for travel and warm-up in the morning. We anticipated that the first hour in each 3-hour block would be mainly setting up, getting everybody comfortable, and sound checking, and the last 30 minutes would be listening back and packing down before the next act. That left around 90 minutes for actual tracking, in which we could record more than one song if the artist was rehearsed. In order to offer the artists more from their involvement we could gift them some Abbey Road production time by agreeing that the first song they recorded would be for our project, but they could use the remaining time to track another song with the same equipment and engineers. They would get to keep the second recording as a multitrack ProTools session to do with whatever they wanted, with no strings attached. That is, they would fully own the copyright in the sound recording and be allowed to exploit it however they wished whilst keeping any monies generated. In addition to this we arranged for the sessions to be filmed by a small video production company, Commuter Films (2021), funded by the University. This company was set up by graduate music production students who understood the recording environment and worked with us to film in a guerrilla style that would both complement the authentic aesthetic of the audio and have a minimal impact on the recordings. The fact that the sessions would have video as well as audio output would be beneficial to the artists who would get synced, edited, and graded footage of their Abbey Road performances as well as the audio. In making these adjustments to the timetabling and the planning process, the motivations of the participants are affected by allowing the artists more control over the time they have in the studio and more control over the aesthetic of their work. The nature of the motivation is important here, as the arrangement should stimulate intrinsic motivation and provide an opportunity for the artists to engage in their harmonious passions (Amabile and Pillemer 2012; Vallerand *et al.* 2003). The offer would likely affect the preparation efforts of the artists, and it projects a positive image of the producers and the institutions partnered in the project. It tints the Production Habitus with a notion of camaraderie and goodwill, which would hopefully become embedded in the final product.

These arrangements could be quite significant for independent artists used to funding their own productions and balancing the risk of it not selling with their drive to make original work. From a systems model perspective, artists are creatively constrained by the rules of the domain and the expectations of the field, which implies that when self-funding a production, higher levels of originality can equate to financial risk. Bennett describes this as a paradox for the songwriter "who is trying to create an original work in a highly evolved, market-driven, and tightly constrained creative palette" (2011, para. 1). The design of our project shifted most of the financial and commercial pressures away from each artist and gathered it into a collective pool where the institutional partners underwrite the overall production costs, and all the contributors share a common interest in simply making the best product they can for the aesthetic and charitable goals. The artists' contributions become all about their performances, and they have none of the usual pressures of post-production, product design, manufacture, promotion, or sales to worry about. In this way the design of the Production Habitus is balancing pressures and removing concerns in a way that allows the artists the freedom to focus on their performance.

Therefore, in our initial communications with the artists we would need to be very clear about the plans; who was getting what, how the project was being funded, licensing and ownership rights, the involvement of the charity and the university, and the level of commitment we were looking for, in order to foster a Production Habitus that would stimulate the appropriate behaviours, motivations, and creative responses. This first communication would also be crucial in addressing a significant concern with this sort of 'various artists' project, which is: how can we combine 12 diverse artists onto an album and still give that album a coherent aesthetic personality? The studio set-up and static backline would be a sonic factor here, but more importantly the project's core charitable aims would inform the themes and messages expressed in the song selection and performances.

Music Minds Matter (see www.musicmindsmatter.org.uk 2020) is a mental health support service dedicated to helping anyone in the music industry who is experiencing tough times. It was launched as a helpline by the charity Help Musicians UK in December 2017, just a few months before the studio dates of this project and was selected as a cause that everybody on the project could feel closely connected to. The effect of the presence of the charity in the

Production Habitus is exemplified by Chris Sheehan, coproducer, in a press release about the album:

> Musicians are the first people to stand up and raise money for charity, even when they're struggling to make ends meet themselves. But who raises for them? We felt it was about time we did something for those musicians who need the help that this remarkable charity gives (…) We passionately believe in the work Help Musicians UK does for our musical family, and we hope people out there will join us in raising money for the charity's vital service Music Minds Matter, which has become part of the album's title.
>
> (Chris Sheehan, helpmusicians.org.uk 2019)

The powerful message of camaraderie and 'musicians helping musicians' was threaded through conversations and communications around the project, and so affected the Production Habitus profoundly. The mental health support concept became a theme that directed the selection of tracks for the album. The artists would be briefed on the charity and asked to contribute a song that was somehow related to going through or getting over challenging times. We felt it a strong enough concept to define the album, but loose enough to be open to interpretation by the artists. Many songwriters would have songs that would fit, and in writing a new song or going through their catalogue to select an appropriate song to contribute, the artists would be undergoing creative and psychological processes that embedded them in the Production Habitus and connected them to the project goals. This would further ensure aesthetic and thematic coherence across the tracks.

A benevolent label, Monks Road Records, was also approached and they agreed to fund the mixing, mastering, and manufacture of the vinyl release of the album. The label's presence on the project and their support for the artists involved and charitable cause reinforced the emerging Production Habitus and its defining themes of social justice, and would influence participants' contributions and motivate them in three key ways. First, the commitment to release the album on vinyl defines the product that will emerge from the Production Habitus in a concrete manner that is easily visualised. It will be an enduring artefact that all artists will be part of – a momentary collective preserved in a physical object. Second, the vinyl format supports the heritage and audiophile aesthetic we were going for. It taps into domain awareness of independent labels and artistic authenticity. However, pressing music onto vinyl is often prohibitively expensive and financially risky to produce for self-funding artists, meaning the format itself lends quality to the project and provides additional motivation to be involved. Third, the financial commitment from Monks Road demonstrates a confidence in the project that will reassure and motivate the artists.

We were a little cautious in how to allocate meaningful roles to the students without putting off the artists. With no disrespect to the students, the artists would want to feel confident that their Abbey Road performance, carefully composed, invested in and rehearsed, would be recorded by professional Abbey Road engineers and not be hampered by the input of less experienced student engineers. The solution was to employ the students as artist liaisons. Each artist would be allocated a liaison who would be their first point of contact leading up to the session and on the day. This is a role often found in music festival environments where multiple artists need to be guided through initially confusing locations and arrangements, introduced to people they would be working with, and made aware of what's going on. The artist liaison is a first point of contact, smoothing the way for the artist, making them feel informed and comfortable, removing obstacles to creativity and addressing any queries that arise. As the specifics of the liaison responsibilities were being devised the importance of the role grew and the potential effect it could have on performances became apparent.

The liaison would be the main point of contact between project and artist, and therefore largely responsible for how the artist perceived the project. How they thought and felt about the project after communications with their liaison would affect their commitment, creativity and behaviour in preparing for the production and their performance on the day. The Production Habitus could therefore be largely defined in those liaison–artist relationships.

This meant the liaisons would need an in-depth understanding of the project to be able to communicate the themes, plans, and agreements effectively. They would need training and guidance in order to portray an appropriately professional but friendly tone in their emails, and instruction on how to respond to certain enquiries. Miscommunication could alter the Production Habitus from how it was intended and result in less appropriate responses and performances from the artists. Therefore, the liaisons were briefed accordingly in a classroom setting, with the roles divided into pre-production communications and on-the-day production assistance. A rough schedule was proposed for rolling out the pre-production communications, and guidance given for looking after the artists on their studio day.

Creating the Production Habitus: First Contact

With most of the plans in place a selection of potential artists was agreed upon and they were invited to take part with the initial contact coming from coproducer Chris Sheehan as he was well known to many of them already. These relationships were vital for the artists' first impression of the Production Habitus and represented a core part of Chris' role on the project, leveraging his social and symbolic capital in the field of independent music that the artists were being drawn from. They know him to be trustworthy, and that he has a reputation for looking after artists' rights and striving to create better conditions in the music industry on their behalf. The existing positive personal connection the artists had with Chris was a foundational building block of the Production Habitus. It was a package of trust and goodwill that would positively taint the artists' experience of the project whenever they engaged with it. The pitch Chris brought to the artists for this first contact was carefully agreed beforehand to paint the project with the desired aesthetic, including highlights of the plans to date. A key phrase used to communicate the production aesthetic and logistical arrangements was 'live at Abbey Road'. This phrase would hopefully trigger an array of ideas and interpretations, drawing on the artist's existing Abbey Road domain knowledge. It has a linguistic capital, a certain cultural value in the field of recorded music that participants would respond to (Thompson, in Bourdieu 1991, p.18). They would visualise the iconic studio, recall the many 'Live at Abbey Road' videos that have been produced with well-known artists, not least from the Channel 4 *Live From Abbey Road* television series (2007), and mentally place themselves within the scene. We hoped the phrase would conjure ideas of quality, heritage, great acoustics, vintage mics and instruments, and the highest recording expertise available. Another key pitch phrase imbued with symbolic power that we used was 'think *Later with Jools Holland* but without the audience'. This could elicit similar ideas of musical quality in live performance but with an added implication that the acts have been carefully curated and selected because the producers believe in their talent. We hoped this would motivate the artists by communicating both our expectations of them, and our confidence in them.

When enough artists had expressed an interest a Memorandum of Understanding (MOU) was circulated as a semi-formal document amongst all parties. This document was a very important part of establishing the Production Habitus, gathering together in one place the aesthetic and logistical plans, as well as outlining the goals for each party and the commitment everybody was agreeing to. It defined the end product, the 'live at Abbey Road' festival approach, and the agreement to give the artists the opportunity to record a second track for themselves. As part of the aesthetic planning, the MOU included reference to the famous heritage keyboard instruments available at Abbey Road, such as the Mrs Mills piano, the Challen upright, the Hammond organ, and the famous Steinway grand, in order to encourage artists to consider using them in their arrangements and therefore bolster the desired aesthetic. In these ways the MOU served not just as a project definition document, but as a blueprint for the Production Habitus in its first physical form. It set the social and stylistic tone for the project, outlined the logistical arrangements, and established structures of capital and motivation that would affect the collaborators' approaches to being involved. Follow-up communications throughout pre-production developed this tone to encourage the intended behaviours and performances in an effort to imprint a sense of the Production Habitus in the produced audio. We hoped artists would prepare well and bring their best game, that they would be excited to be involved but feel relaxed that everything was well organised and in control. We wanted their arrangements to consider the Abbey Road environment and to favour organic tones and textures that made the most of the studio's facilities and instruments, keeping to ideas around authenticity and of brilliant performances being captured in the best way possible that are evoked by the cultural capital in the name 'Abbey Road'. We hoped their performances would be soulful, inspired by the history of the studio and the themes of the album. Our communications used phrases such as 'live at Abbey Road', 'musicians helping musicians', 'beautifully maintained heritage instruments', and 'recording at the most famous studio in the world' to foster these ambitions. In short, the MOU and pre-production communications were designed to make everybody feel well-informed, valued, looked after, and creatively motivated.

Preserving Production Habitus in Pre-Production

The plan was to have one main point of contact between the artist and the project running up to the session, and this was to be the liaison. However, despite the carefully arranged sequence of emails the liaisons sent to introduce themselves, gather technical riders, and distribute logistical updates, the artists often tended to bypass them and approach the producers Chris Johnson and Chris Sheehan directly. This was largely down to many of the artists already knowing the producers personally or being introduced during the MOU stage of the project and defaulting to their more familiar communication channels. This is understandable given the students' low level of cultural capital. It is possible the liaison role in pre-production was viewed as artificial, as though part of the cost of being involved was to go along with the student-as-liaison scenario whilst seeking more confident communications from the coproducers. The liaisons collected technical riders and line-up information from most acts, but as the dates of the studio sessions approached, the coproducers took on more of the artist communications. It became clear that this was needed to preserve the Production Habitus as artists sought quick and reliable responses to any questions that arose, and the coproducers were best placed to remove any potential stressors and logistical obstacles. The presence of the liaisons during this process was still important in setting a tone for the Production Habitus. The tight organisation and the social and educational goals created structures that the artists were aware of that would still influence their perception of the project. It arguably established a more robust communication network and facilitated a coherent Production Habitus, so that every time the artist engaged with the project in pre-production a consistent tone was presented.

A challenging aspect of the Production Habitus that needed to be addressed in pre-production was to control the flow of people through the studio, managing the expectations of the participants in advance. This would include considerations to do with the physical environment, the social environment, and the aural environment of the recording sessions. With the students and production team already based in the control room it could quickly get out of hand if artists invited guests or a full entourage to witness their Abbey Road session. The control room is not a big space to have more than a few people in, and the presence of others not fully involved in the production can quickly lead to unhelpful levels of chatter that distract the producers and artists. In this project, there was also an educational factor whereby the students needed to be able to observe the Abbey Road engineers, which required a certain amount of space to move around the studio and a lot of concentration that is easily lost with the presence of visitors. Most importantly, the distraction to the artists themselves would be unacceptable if every time they heard the control room environment over the talkback mic, it sounded noisy and indifferent to their efforts. However, a certain amount of traffic through the control room would be unavoidable given the multi-stakeholder design of the project and the need to keep significant influencers excited about it and personally connected to it so they would support post-production and promotion efforts going forward. Significant guests from Help Musicians UK, York St John University, Karousel Music, and Monks Road Records would be welcome in the control room due to the capital they were bringing to the project. Additionally, journalists and other key influencers who could help promote the project would need access to get on board with what we were trying to achieve. It represented a challenge to be able to invite people into the bubble of the project, our carefully designed and maintained Production Habitus, without their very presence upsetting it. For these reasons the pre-production communications with the artists set up a rule and a plea for understanding that they should only bring musicians who would be playing with them, and that we were unable to host friends, partners, managers, and other guests.

The late stages of pre-production added the Abbey Road engineering team to the collaboration, and they were brought into the Production Habitus by an initial conversation with the lead engineer, Paul Pritchard, in which all the aims and concepts of the project were communicated. It was imperative that the engineers had a full understanding of the Production Habitus at this point as they would play a major part in rolling it out on the studio days. Paul and his team would be instrumental in designing and controlling the Physical Environment and the flow of activity through the studio days to match the Production Model and our aesthetic intentions. They would play a major role in managing the overall stress and comfort levels of the artists in the manner in which they communicated with them during set up and their performances, thereby affecting the 'Persons' element in our model (see Figure 3.1, Chapter 3). Important parts of the project's Social Micro-Culture during the studio sessions would be defined and communicated through the engineers' behaviours and conversations. Therefore, following the initial conversation on Project Definitions, in-depth discussions took place between Paul and I to design a studio layout that would achieve the acoustic and logistical goals whilst stimulating the creative energy, social chemistry, and considered performances that we hoped would define the project's aesthetic.

Production Habitus in the Studio Sessions

The night before the session, Paul and his team began the studio set up. The key factors in arranging the space were the live performance Production Model, the need for quick changeovers, and the desire to keep all the musicians comfortable and happy, removing any obstacles that might keep them from performing their best. Line-up and technical information had been collected from each artist and these were considered in designing the layout of the studio floor in a way that would suit each act with the minimal amount of adjustment between each one. Monitoring and sightlines were considered paramount, alongside the need for isolating the drums from the main room and baffling amps and keyboards where necessary. The drums would be set up in the largest booth but with clear sightlines through the booth's glass doors to the area where the bass players and singers would be in the main room. The main live room was effectively halved so that the space the musicians would play in would keep them quite near to each other, at a proximity they would be used to, to elicit coherence in how they play off each other as a unit. We felt the separation we could achieve in the large space by spreading them out across it would be felt too much in the performances, and that keeping to dimensions more similar to a large festival stage would provoke a comfortable performance atmosphere more conducive to expressive, confident and focused performances.

Free-standing baffles were positioned in a loose semi-circle that would be the back of the singing position, and also the effective limit of the performance space at halfway along the studio floor. The desire for an open space with clear sightlines meant we didn't want to fully enclose the vocalists, but we needed to control the acoustic effects from the large room entering the live side of the cardioid vocal mic, as well as baffling the grand piano. A keyboard station was assembled next to the singing position, with the Steinway grand piano, the Hammond organ, the Fender Rhodes, and a Nord stage piano in a cluster with each instrument mic'd or DI'd ready for use. The Leslie cabinet was positioned way back in the room using the built-in wall baffles to create an enclosure and give it some isolation. Bass players would play in front of the drum booth with clear sightlines to the drummer and the

Figure 4.1 The studio floor layout. Only half the floor is used with clear sightlines between everyone. The guitar amps are boxed in the cubic row of baffles behind the guitarist. The bass position is bottom left, marked by the personal monitor mixer, and the drums can't be seen here, in the booth off to the bottom left. In this picture are Romeo Stodart on guitar, Ren Harvieu at the vocal mic, with Luke Wiggs and Adam Townsend of Commuter Films, and Dom Smith, session photographer. Much more detail can be seen in the videos produced in the session, available at https://tinyurl.com/y426fnc6 or search YouTube for *York St John Music Production at Abbey Road: Smoke Rainbows*

singer, with the bass amp in the smaller booth. Small baffle enclosures were built in a row to isolate a selection of classic guitar amps giving guitarists a choice of tones with minimal set-up changes between acts. The guitarist playing position was in front of them. An area of space was left open opposite for artists that were bringing string sections, and the chairs, music stands, and monitor feeds would be ready to put into position for them.

Each playing position had a personal monitor mixer, fed by sub mixed stems from the control room. This allowed each musician control over their headphone mix without having to keep asking the engineers to make adjustments, which was an essential factor in facilitating the rapid set-up required by our production model. Williams (2012) has pointed out that personal monitoring systems like this remove a potential source of tension between musician and engineer, and it is another area in the Production Habitus where a sense of Ownership becomes important. The musician is not reliant on the engineer in shaping the audioscape they will inhabit for the performance but can control it to a large extent independently without disrupting other tasks going on around them. However, Williams also argues that such systems can be detrimental in fine-tuning arrangements, as a musician who hears an offending note in another's performance can deal with it without conversation and risk of conflict by turning them down in their individual mix, thus leaving the problem undiscussed and unresolved. For our project, we were relying on all such refining work to have been done by each artist in their preparations as a result of the Production Habitus created around the project and their responses to it. The inclusion of the personal monitoring system also addressed some of the considerations in the 'Persons' element of the Production Habitus by considering the social and psychological state of the performers. Being able to quietly control their personal headphone mix removed the social stress of publicly requesting a lot of minor adjustments that may be needed to feel comfortable, but which the very act of asking for could be perceived as fussy. It also facilitated the musician in creating their own audio space to disappear into in the midst of a busy set-up and sound check. A private audio 'backstage' where they can mute everyone else and quietly run through their part, doing the creative and psychological preparation needed to 'get into the zone' before the recorded performance.

This basic studio design considered the technical requirements of all the artists and could be quickly adapted to accommodate them in the short turnarounds between recording slots.

With all the elements for the Production Habitus carefully considered, designed, and assembled the coproducers' role in the studio sessions became more hosting and stage management to ensure everybody was introduced, felt comfortable, and knew what was going on. The liaison personnel assisted here, meeting their artists on arrival and showing them the studio before loading their gear directly onto the studio floor where they would be introduced to the lead engineer, Paul, and shown their playing positions so they could make themselves comfortable. They were given a quick tour of the studio and shown the shared green room area at the back of the control room, a single room where everybody could leave their coats and bags or have conversations without disrupting the session. The Production Habitus was continually reinforced in the small talk as participants meeting for the first time discussed the project, their part in it, and shared positive reflections on the project and its potential impact. The PR effort of the project began in earnest in the studio sessions, and everybody was encouraged to post about the sessions on their social media pages. It was recognised that social media posts on the more well-known artists channels would add value to the project, and that other collaborators could gain cultural capital in this new association. This aspect was viewed very positively, as the project's commercial and social success would benefit all involved as well as the causes of social justice that they support. In accordance with the sense of camaraderie baked into the Production Habitus, the tagging

Figure 4.2 An adapted set-up, with the chairs for a string section visible at the bottom of the image. Note also the addition of a three-sided booth on the left, assembled here for a saxophonist

details of the partners and the charity were printed out and displayed in the control room so that everybody's posts could be mutually beneficial.

The effect of the Production Habitus was evident in the preparation most of the artists had put into their contributions. They were incredibly well rehearsed and had gone to great lengths to arrange their songs specifically for the project, adding live string sections or hiring additional session musicians where necessary. They were respectful of their time slot and keen not to overrun, were graceful and amiable in their social conduct, and had selected songs appropriate to the mental health concept of the project. It is interesting to examine the one exception to this to interrogate what might have been different in the way they interpreted the Production Habitus.

We had attempted to control the numbers of people in the control room by explaining to artists beforehand the limited space and the need to give the students and engineers the chance to concentrate. Most artists understood and only brought musicians with them; however, one artist arrived with a large entourage and set up camp in the green room in a way contrary to our spirit of camaraderie. The same artist had not prepared a live arrangement in the way we had hoped, augmenting live drums and guitar with a heavily layered prerecorded track for many of the backing vocal, percussion and synth elements. The acoustic version they had prepared as an alternative was a simple guitar and vocal arrangement that we felt missed the aesthetic guidance we had been communicating, failing to make use of heritage instruments and acoustical opportunities the studio afforded. Similarly, this was the only artist to push the limits of the allocated time slot by continuing to run several songs after several versions of the two agreed had been completed. When this became apparent, the presence of their manager made it a little more difficult to intervene, setting up a communication structure that demanded a firmer tone that was outside the Production Habitus we wanted to maintain.

In analysing why this artist did not display quite the behaviours and attitudes we had tried to elicit, and why they had seemingly missed so much of the essence of the Production Habitus, we can suggest it was because they had been recruited to the project via a slightly different route. Most of the artists were carefully curated by the producers and were connected to Karousel, the not-for-profit musical collective managed by Chris Sheehan. In selecting the artists, Chris considered the moment they were at in their career and how they would approach the Abbey Road project opportunity. Some were early career independent artists who had built the right level of momentum to be able to perform at a high level, but who had enough experience to recognise what being part of the project could mean for their careers. Some were at a later stage in their career and would appreciate the unexpected chance to record at Abbey Road, perhaps when they had never had the chance to earlier in their career. As most artists were connected to Karousel they were also familiar with its community spirit and of being in situations where multiple artists were in shared situations, needing to help each other out for the benefit of everyone. These artists approached the project with a reverence for the studio and an enthusiasm to be part of a multi-artist album with its stated aims. Our errant artist was unknown to the producers and included at the request of the charity who were using one of their songs as the anthem to launch the Music Minds Matter campaign. It therefore made sense that a version of the song feature on the album we were producing. However, this alternative path to the project meant that the artist was not subject to the same considerations as the others in selection. The line of communication between the producers and the artist was longer, and so messages about the nature of the project were filtered through a support team surrounding the artist. We can postulate that much of the social and conceptual frameworks of the Production Habitus were lost in this chain of communication. As this was a relatively new artist tipped for stardom, their career was at a moment where their support team would be arranging a lot of opportunities and exposure for them and giving the artist a lot of encouragement to build their confidence. In this context, with the artist's head being filled with confidence-building messages from their team, our project could be viewed primarily as another opportunity that they rightfully deserved, with our actual project ethos and aims largely overlooked. They brought with them their own existing Production Habitus, with a social micro-culture supported by the team who arrived with them. This effectively acted as a bubble that shielded them from the effects of our Production Habitus, and as such, they approached the project with less reverence and with less consideration for all the others involved.

This highlights the importance of the producers' role in selecting the 'Persons' for the Production Habitus. It shows by comparison how successful the curation of the other artists was, and how much consideration of their career, dispositions and sensibility is needed to craft a coherent aesthetic. We can also consider the challenges of the communications needed to generate the Social Micro-Culture. Our approach failed to get the right messages across when they needed to pass through a support team who perhaps had an alternative agenda for their artist.

Production Habitus in Mixing

The principal decision around mixing the album was deciding who was going to do it. Early conversations focused on the budget, who we might already know that would be sympathetic to the project aims and so give us a good deal, and who might be available that had a good reputation for mixing records. Our ambition for the audio quality of the mixes was very high.

We felt that given the world-class quality of the recording situation, the mix had a lot to live up to and so we wanted to source a mix engineer with an established reputation and sufficient cultural capital in the field whose involvement would further enhance the actual audio whilst elevating the project's standing in music industry circles. By the time the recording sessions came around we hadn't yet selected anyone to mix, but we had a much deeper understanding of the Production Habitus and therefore the selection criteria for the mix engineer had become more nuanced. We had been enjoying a strong working relationship with Paul, the recording engineer in the sessions, and the notion of a 'name' mix engineer became less important. Paul's mixing skills were unknown to us, but he knew the material and project aims intimately having been immersed in the Production Habitus and this had a high value. In the sessions we saw that he was a brilliant recording engineer, and he was a very positive presence in the social microculture. He was a core part of the conversation when we were discussing tonal aesthetics during the sessions, and so it just 'felt right' that he should mix the album. This sense was so strong that it overcame our concerns that as Paul works at Abbey Road, hiring him to mix also meant spending mix budget hiring another Abbey Road studio, the Gatehouse, for a few days. This would shorten the time for mixing to just three days, which we rationalised suited the 'live on the studio floor' aesthetic. The mixing sessions happened a month after the recording sessions and keeping the production at Abbey Road gave a continuity for the producers. Having been away from the studio for a month and being busy with other projects, walking back into the same building to return to this project had the effect of physically putting us back into the Production Habitus. The project aims, the social dynamics, and the intended aesthetics were instantly refreshed and activated in our awareness. The symbolic capital of working at Abbey Road continued to have an effect on us throughout the mixing.

If we consider our Consolidation Points Framework (CPF) from the previous chapter, we can understand that the mixing process combines much more than just audio. With the coproducers in attendance, Paul's task included blending the personalities, meanings, and moods expressed in the performances with the project's mental health message, its defined aesthetic goals, the acoustic sound of Abbey Road Studio 2, and the comments from the coproducers. We were considering the artists' positions and predicting how they might react to certain mix decisions that foregrounded one instrument over another or that enhanced or diminished the natural acoustic of the space. Conversations would keep referring back to the 'live aesthetic' and the product definitions we had made in constructing the Production Habitus early in the process. The mix needed to stop short of correctional editing or overly processing sounds in order for the 'liveness' to be overtly audible. There was a need to enhance the organic nature of the tones and performances, to keep the occasional shuffle or count-in that signalled it was all recorded live, as well as to produce a vintage warmth and smoothness that would satisfy our conceptions of what an 'Abbey Road' sound should be. We were essentially consulting our domain knowledge, aurally imagining what it should sound like, and then attempting to communicate this image, sometimes with objective technical terminology ("Pull out about 3db at 200Hz and slow down the attack on the 1176"); sometimes with shared domain references ("Drier than *Solid Air* – closer to Joni Mitchell on *Blue* – but keeping the John Martyn warmth") and sometimes with more creative, subjective language ("It needs to sound more yellow and dustier"). We were comparing and negotiating our individual aesthetics through conversations sited within our shared domains of production, songwriting, music technology, and popular music, and shaping the tones of the product toward our coproduced idea. In this way the 'mix' included processing and balancing the nested holons and domains of the coproducers into one aesthetic that the product would embody.

There were also considerations around the sequencing of the album, which led to us selecting the second song recorded for some of the artists, the one they recorded for themselves in the remaining session time rather than the one they had intended for the album.

This meant that we could address sequencing issues where the album felt like it needed more energy at certain points, or shorter songs at others, and it enabled us to avoid the inherent risk of producing an overly downbeat album due to the mental health subject matter. Of course, this was done with the artists' consent, and it allowed us to address a limitation in the project planning, which had meant that we didn't hear demo versions of all the songs before the recording sessions. The short turnaround on the mixing still allowed time for the artists to comment on the first mix, requesting amendments before committing to the final mixes. In this way they could remain part of the Production Habitus, able to contribute to post-production. It is interesting to note that most artists were considerate and kept their requests simple, mainly addressing broad level changes that reflected intentions in the arrangements that we had missed. I suggest doxa from the Production Habitus to do with camaraderie, a shared understanding of the multi-artist project, not wanting to be overly precious in this situation, and knowing the project's time and budgetary limitations prevented artists from commenting as much as they would have done in projects where the creative ownership was more with them.

Production Habitus in the Product Preparation

The project's social aims came to the fore in preparing the actual product, with the main focus being on the vinyl version of the album. Copyright issues prohibited us from using the phrase "Live at Abbey Road" in the title, and many potential names that would summarise the mental health aims of the project in a short snappy phrase were floated and discussed. The name needed to balance these social aims with fresh language that would be appealing to a popular music audience. Most of our suggestions were dismissed for sounding overly 'worthy' or just not interesting enough, until Chris Sheehan recalled a lyric from back when he was a songwriter that used the phrase 'Smoke Rainbows'. He explains in the liner notes what that meant:

> Writing songs was my way of turning darker times back to colour so that it might help both myself, friends and strangers; but eventually the colours ran out. It's often only human warmth and contact that brings us back from the monochrome.
>
> (Chris Sheehan 2019, *Smoke Rainbows* liner notes)

This concept of a desaturated rainbow was communicated to Scott Jones, the art director for the project who had been attached from the beginning and so was immersed in the Production Habitus and fully aware of its aesthetic and social aims. As the album was aiming to raise funds and awareness of the Music Minds Matter mental health service for musicians, Scott wanted to extend the musicians' collaboration into the visual artwork. Each artist would draw or paint a line on paper that represented their life in music to date. Scott would then combine these lines into a sort of collective consciousness that became a 'rainbow' for the album cover art. Using heat sensitive ink, this 'rainbow' appears completely black until you hold your hand on it, whereupon the body heat melts the black away to reveal the colourful collage of lines underneath. The idea being portrayed is that darkness associated with mental health issues can be lifted with human contact. This was an unanticipated but brilliant demonstration of one of the core research interests of the project, which was to investigate how the essence of a Production Habitus can be rendered in the final product. The visual interpretation of the project's social aims and the camaraderie of the social micro-culture are blended with the personalities, histories, and creativity of all the participants into an interactive physical artwork. Notions of high quality that we built into the Production Habitus are also present in the rest of the graphic design package, where the back of the record sleeve is pure white with a simple message in an elegant font printed in the centre: 'Music Minds Matter.' The organic, classic recording aesthetic of the studio is represented on the inner sleeve, which features photos from the sessions processed with a vintage tone and subtle film scratch filter.

This project was perhaps unusual in its particular multi-stakeholder, multi-artist dimension. However, by using the concept of Production Habitus as proposed in Chapter 3 it was possible to manage the detailed planning needed to create a coherent set of goals to mobilise the participants and to produce a coherent product at the end. By considering the project definitions first and then the motivations of the possible participants we were able to build a production process that enabled creative and appropriate responses. Drawing on and merging our own domain knowledge as musicians and producers we were able to use language and communications to define a social micro-culture and a way of doing things that the collaborators could buy into. We were sensitive to the capital in the studio, the stakeholder institutions and in the people working on the project and this allowed us to shape the feel of being involved in the project so that we could attempt to stimulate desired behaviours and encourage those ideas that worked with the project goals. As a model for designing and for thinking about a coproduced project, Production Habitus draws creative attention to important components that may have otherwise gone overlooked or undervalued.

5 Lauren Christy and The Matrix Production Team

Coproduction in Familial Mode (the Three-headed Monster and the Butterfly Collector)

Robert Wilsmore

The Matrix are a songwriting and production team that formed in 1999 and had their first major success co-writing and coproducing Avril Lavigne's song 'Complicated' in 2002, winning the Ivor Novello Award for International Hit of the Year 2003. Amongst their list of accolades, they have also won the BMI and ASCP Songwriters of the Year, they are twice winners of the Canadian Juno awards for Pop Song of the Year and Album of the Year, and are Grammy nominated seven times. They have had:

> Numerous songs in the US Top 20 Billboard Charts in various formats that they co-authored and produced and held the #1 spot for over 19 weeks. In total, The Matrix's songs have sold over 35 million records through the various artists they have worked with.
> (thematrix.com 2018)

Around the turn of the year 2000 The Matrix embodied a new direction with which to start a new millennium. If, in a thousand years, someone wanted to study a production team at that moment in time, the beginning of this thousand-year block, then The Matrix might well be their choice of case study. They formed with the name The Matrix before Baudrillardian simulacra took to the screen in the form of Keanu Reeves as Neo in the film *The Matrix* (1999), which captured the zeitgeist of a society struggling with reality. Conversely, what the production team did was to shift us away from a dystopian vision of the future. They captured emotion and aggression in music from themselves and the artists they worked with, bringing the authenticity of humanity to the fore. It was not a millennium starting with the perils of artificial intelligence; their music was human, first and foremost, human. Reconnecting us with ourselves and our senses in the wave of *affect* that was taking place at the time and of which they were a part. The term 'the matrix', for this team, referred more closely to the Latin 'mother' (mater) and more pertinently to the 'womb', a place where life is brought into being. The film, to which the name is now attached more firmly in society, has the womb as the place of actuality but where that actuality is dystopian and abused (where people are energy cells for aliens). Thankfully, the production team of this name draws us back to the matrix, the womb, as essentially a place of humanity, a place of growth and becoming.

Behind the name are three artists: Lauren Christy, Graham Edwards, and Scott Spock. Christy is an England-born singer songwriter who signed to EMI Mercury Records in her teens and released two albums, *Lauren Christy* (1994) and *Breed* (1997). Edwards, an England-born musician, was a professional bass player who in the 1980s had played for many well-known artists including Adam and the Ants and toured for four years with Go

DOI: 10.4324/9781351111959-6

West before becoming, as he writes, "tired of being on the road, so I started getting into songwriting" (Kawashima 2006). Christy and Edwards were husband and wife at the time The Matrix formed and for the period in which they were so successful. Spock, an American from St Louis, America, played jazz trumpet at school and had a degree in Jazz Arranging but "realised there was no money in being a trumpet player [so] I got into the whole computer side of arranging" (Buskin 2006). Graham came across Spock when the latter mixed a song for Graham's Los Angeles group DollsHead and asked him to join the band. The band's manager, Sandy Robertson, suggested they worked together writing for other artists. For the team, a combination of being tired of touring, of being the frontline artist, of record deals coming to an end, of realising that husbands and wives can work together, of 'clicking' as a working group, of being three and not two, of being in a similar position and similar mind to work 'behind' the artists, all made for the right combination and environment. Three talented musicians and artists with complementary and shared skills, with a shared and transformative vision, made for a positive alignment that proved highly successful.

As a production team they have strength and depth in all the areas of music production so that they can overlap roles, change roles, and offer skills that are complementary to each other. They hold aspects of familial and integrated models of coproduction that are key to their success. This fluidity of roles and tacit permissions amongst the group to allow contributions to be made to the project (until there is necessarily consensus or a two-thirds majority decision in this case) ensure that the product is where they want it to be before it is released. And, although popular success was a goal, it did not override the need to produce what *they* want to produce (*their* shared aims and vision). They have specific skills, shared skills, a shared vision, and a transformative vision. They transformed a talented young singer from an uncomfortable Faith Hill-styled artist into an authentic Avril Lavigne. The Matrix had that transformative skill to make the 'talented' become the 'artist'. Where this ability may have stalled at times (as happened with Katy Perry) it was not necessarily down to them, but down to how others wished to *forefront* the team rather than let them forefront others, a decision that clashed with their shared aims and vision. With their backgrounds as artists, as performers, they have insider knowledge to draw on. To reiterate, Edwards having played bass for many renowned artists in the '80s eventually tired of touring and found production more appealing, Christy was a signed artist having released successful solo albums but wanted to move behind the scenes, and Spock was a trumpet player with a degree in jazz arrangement who found himself drawn to technology as it had a more relevant role for him and for society than his trumpet playing might have had. They all moved from their diverse professions towards being one production company that brought all those skills together.

In an interview with Richard Burgess, Christy recalls that The Matrix had their 'big break' with the invitation, through their manager Sandy Robertson, from A&R executive Ron Fair to write for a Christmas album for Christina Aguilera. They wrote the song 'This Year' from the Album *My Kind of Christmas* (2000) and "Rob Fair loved it and said that he would co-produce with us for the record." Burgess asks the question "Was he in the studio with you a lot?" The answer to which is "not at all". Partly due to time constraints that meant that it was not going to be possible, Fair, who was on tour with Aguilera at the time, had her vocals recorded during that time using a portable studio. Christy, Edwards, and Spock worked on the track and "Then the three of us got together with Dave Pensado and Ron and mixed it. It was a real co-production" (Burgess 2004, p.2). Although it would be easy to attach too much significance to the phrase "a real co-production" here, it presents an opportunity to see how the term is viewed through the various coming-togethers of different producers and production elements. The three producers in the team The Matrix are already a 'real' coproduction team, but because there are other producers not part of that (one) team that are in the production of the music, namely Ron Fair and Dave Pensado here, then the term 'co-production' is used, the 'real' just emphasises the bringing together of others outside of The Matrix (the three-as-one) in this, the making of this record. The Matrix productions are already coproductions, that is, they are the work of three people, Cristy, Edwards, and Spock, and not one single person. However, in this case study The Matrix form a singular identity; songs are 'produced' by The Matrix, not 'coproduced' by The Matrix, that is unless someone outside the group, as is the case in Aguilera's 'This Year', are also part of the production process. The term coproduction here applies to the group that is The Matrix + Pensado + Fair + Artist. We could turn it into an equation where coproduction (cp) is the sum of The Matrix (tm) and Others (o). Hence cp=tm+o where tm=c+e+s (Christy+Edwards+Spock). Hence cp=c+e+s+p+f+a, or cp=tma+o. This rather awkward example is merely to express how groups can be identified as ones or as 'ones that are collections of ones' depending upon how we group them. And in this case, it highlights how three are 'one', and adding others to this one makes the 'co' in coproduction. It is all a matter of how 'one' is perceived. And besides, The Matrix appear generous in their credits, co-writing and coproduction credits often include the artist, as is the case with the writing of 'Complicated', which is credited as being by Christy, Alspach, Edwards, and Lavigne.

The Matrix as a collective have characteristics of 'the one *and* the many', and this comes through in how they operate. "We're kind of like a monster with three heads", says Christy (Burgess 2004, p.4). The Chimera in Greek mythology is one beast but with three heads, those of a lion, a snake, and a goat, and in general terms represents the coming together of disparate (dissimilar but complementary) elements to make a whole, usually alluding to the sum of the parts being more than the individuals. If taken separately, the lion is likely to eat the goat and the snake is probably going to hide under a nearby rock at that point but brought together with the same aims and as allies their complementarity increases their success. In the case of The Matrix there are a number of elements that align; they have complementary skills but also overlapping skills and shared values. In an interview in *Sound on Sound* with Richard Buskin, Spock and Christy said that:

> "As we come from such different musical backgrounds, we approach the music differently, yet we always end up in the same place," Spock explains. "I bring my musician's sensibility to whatever I do. Graham approaches things very organically, from a guitar and bass point of view, and Lauren develops melodic ideas on the keyboard. [...] "Our roles overlap," Lauren Christy adds. "We all have things that we're really, really strong at, but we also involve ourselves in all aspects.
>
> (Buskin 2006, pp.141–142)

Their collaborative approach maps well to that of Vera John-Steiner's 'family' model and hence our familial mode of operation whereby trust and common vision are central in this type. Coproducers will be highly 'familiar' with each other, will have become used to each other's ways of working, and cross over roles easily. In this mode any division of labour might change and expertise becomes dynamic. This is a 'pre-Integrative' model where there are still identifiable contributions but much crossover. Here one might find coproducers feeding in and out ideas and activity in all stages of capture, arrangement and performance, comfortable enough to know that one has permission (indeed encouragement) to do so from the 'family'. One observation is that, at this time, Christy and Edwards were married and as such were literally 'family', but for many years they deliberately avoided working together up until the point where they realised that they *could* work together without detriment to the relationship (Buskin 2006). Scott Spock as the other party then operated to mediate any potential conflict between the two. Already within the 'one' of The Matrix there is the 'two plus one' as well as the three individuals and the three-as-one. The complexity of layers of groups and individuals within groups-as-individuals demonstrates the richness, the depth, of this team, and that these layers align in a positive fashion is evident in the final product. In an interview in *Music World*, Rob Patterson writes that:

> The collaboration among the three "is like playing tennis with someone on the other side of the net who's really good," likens Christy. "When the three of us come together, because we are all so different, we don't really butt heads," notes Spock. "We all respect one another, and once we reach that middle ground on something we are working on together, we know we are onto something good. It's like a family thing when we are in a room together".
>
> (Patterson 2009)

The team note their differences and their similarities. In keeping with the chimera analogy, that of three distinctly different heads but one body, Christy articulates how they may seem to inhabit separate roles to the observer at first sight, but that this is not the whole truth:

> If people come around they'll see us and they see Scott sitting in front of a whole lot of gear and they see Graham with a guitar and they see me sitting with a [note] book. They immediately go, 'oh that's what they do.' Truth is, the lines are so blurred [...] we'll all just blur each other's lines, I guess that's what makes the partnership great.
>
> (Burgess 2004, p.4)

This makes sense in a team where permission to 'blur the lines' is welcomed. We have noted previously in discussions on George Martin's recollections at Abbey Road that division of labour in the studio was such that any crossover might be a case for the unions to have to consider (Martin 1979). But here it is the crossover of roles that makes the collaboration work. Where there is division, then it is helped by one of the key characteristics of a trio which is that there is no 50–50, no binary opposition, or at least not one that produces an equal balance. The equal tug of a 50-50 split is not possible here but rather complete consensus or a two-thirds majority verdict wins out.

Another characteristic of their approach is the ability to zoom in and out of the mix, to focus on details (the many) and to view the whole mix (the one). In a previous article exploring the mix as one and multiple, 'The mix is. The mix is not' (Wilsmore and Johnson 2017), we explored three types of how the mix is perceived, drawing on concepts by philosophers Alain Badiou and Deleuze and Guattari (the latter two being 'one'). Badiou's two forms of the multiple, the 'consistent' multiple, has the mix that consists of 'ones', in this way the listener can hear the elements within that make up the whole and will make judgements based on that approach. But the 'inconsistent' multiple does not have any such 'ones', it is 'pure multiple' (it is *in*-consistent because it does not 'consist' (Badiou 2007)). In music production terms, the listener in this mode hears only 'the one'. It is probably fair to say that listeners have a preferred mode for listening to the recording and will make judgements according to the criteria for that mode. But the production team that can switch modes, or at least can cover all such modes, has the advantage of being able to make judgements on all levels and compare them, and hence are more in control of 'what works' with regard to producing that end product. Christy details their different and complementary listening modes:

> We listen to music so differently. Scott [...] can hear the real intricacies in music like how the hi-hat pattern is fighting with the vocal. Graham [...] will listen and say, 'these

> guitars aren't right, it has an old school feel and doesn't work with the song.' For me, I listen, literally, as a punter. I let the whole thing wash over me.
>
> (Burgess 2004 p.2)

It is after a critical listening process such as this that they are able to go into the production and work on the various levels. The punter is the listener for whom the inconsistent mode of listening is predominant. It is the experiential engager with the recording, the one that hopes to get 'lost in the music', to be void of a conscious and analytical mindset that would disrupt the experience (immersed and in 'flow'). But to achieve that for the punter, the producers have to work on the ones, the elements, so that they allow the final product to be a whole, to be one, to *in-consist*. Ironically then, the producer's goal is for the recording to be *inconsistent*. To use Burgess' own words on one of his types of producer, that of Type C, "If the collaborator had a catchphrase it would most likely be 'the whole is greater than the sum of the parts"' (Burgess 2002, p.7). We might use Badiou's concepts to offer a variation and say that the collaborators motto could be "The whole is the disappearance of the parts". Or even, to go down the pure multiple route, "The one doesn't consist of anything. It's just one", but for a catchphrase it's not very catchy, so maybe it is best to stick with 'the sum is greater than the parts' when it comes to The Matrix.

Their working methods are flexible and reflexive in their actions. Because they share skills, they can deploy the most appropriate group member to suit the situation. In this following example the focus is on the vocal performance of the artist, each production team can work with vocalists, they share that skill, but that can be nuanced according to the situation. Christy gives the example that:

> If it's like a rock thing, it might be really good if Graham does it, whereas Scott's brilliant at working with R&B singers and getting people to be good in the pocket, and sometimes I'm good at working with the younger artists, making them feel more confident if they're really shy.
>
> (Buskin 2006, p.146)

There is no set division of labour within certain categories, there is no one person in charge of vocals, they are all responsible and are all responsive in that they adapt their input according to what will produce the best result. Spock notes the same with regards to the technology, "We're all very versed in terms of how a studio works, and we've got pretty much all the vintage gear you can think of preamp wise – Neves, APIs, etcetera." Spock appears to be the one that runs the 'tech' side of things, but it becomes clear that these are not distinct territories occupied by the three individuals but spaces where they may have a preponderance over the occupation of that area rather than ownership. Hence Spock seems to be the techy-one (and our Trekkie mentality is likely to force some sort of nominal determination here in relation to the techy 'Scotty' and the geeky 'Spock'). At the time of our interview with Phil Harding, he too drew on the analogy that had been made of them at SAW (in a cartoon drawn by one of the artists they produced). which had the comparison of them as the team running the bridge of the *Starship Enterprise* (see Harding 2020). If the three, The Matrix, are reduced (perhaps unfairly) to essences then Scott Spock would be given the techy label, but we can see how tech and artistic practice are *not* separate but are synergistic. Talking with company Softube, Scott explains that:

> Every day we enter the studio, we tend to do different types of music and always try to build in production as we write, you know because we feel that the music is just really one artistic statement, it's not like 'OK, let's write the lyric now, now let's write the

> chords, now let's mix it' you know, it's like the way technology has evolved, has brought us to a point where music tends to evolve organically within a technological aspect within computers and plugins and everything. Your mix starts to happen because you want a specific sound and you want things to react quickly because you're in a creative moment.
>
> (Softube 2010)

Here we are beyond the technology 'capturing' the sound or recording. It is part of the instruments of the musicians *whilst* they create. A tool helps to create what is desired but is generally separate to what is desired; a musical instrument on the other hand is simultaneously a tool of production and part of the desired production, its dual position placing it as both 'instrumental' in terms of being a means to pursue an aim, and with regard to 'being' part of that aim. This dual position is what Spock emphasises here, they "build in the production as we write", and this integrates technology and creativity. What is more, rather than referring to a whole that is created, Spock remarks that "the music is just really one artistic statement", as such it is not a whole made of parts (a consistent mix) but rather it is 'one', an *in*consistent event. This resonates deeply with us, with our 'Song of a Thousand Songs', our *all-songs-are-one-song* song, which can be summed up in that exact phrase "the music is just really one artistic statement". Humankind's artistic statement of itself and to itself.

Similarly, in the evolving process of creativity and tech combined, they do not see that there is a particular formula in the way in which they go about songwriting (a term that for them, as we have just seen, includes 'production'). There is no set pattern to song structure that is adhered to, but they are aware of the various compositional techniques and structural possibilities that they can use. Spock points out that:

> There's no set formula on the way we write songs. All three of us bring ideas to the table and decide what's best. The secret to our success, if anything, is that we are all so different and bring so many things to the table.
>
> (Patterson 2009)

Christy says that "Our dynamic is very specific to having the three of us, our three diverse personalities. It wouldn't work any other way" (Yamaha 2018). In truth it is the combination of being 'all so different' whilst simultaneously being 'all the same' that is key to the success. Difference and repetition. But this must be qualified with regard to what is different and what is the same, particularly as the term 'all the same' sounds somewhat damning, lacking in individuality and hence in creativity and artistry. Where the differences can be established (Graham and rock music, Spock and R&B, etc.) the sameness is that of the same aims, the same transformative vision, the same driving force that are manifest in the words "Our Passion is Music", which is the bold statement at the top of their website (thematrix.com 2018). Spock's comment that "we pride ourselves on capturing emotion and aggression" (Strauss 2002) can be seen across the genres they have worked in from Britney Spears to nu-metal band Korn; it speaks to a consistent theme throughout popular music, that of the cry of expression, the need for the artist to communicate something authentic, something that, if it cannot *only* be done through music, is *best* done through music. Affect has primacy, we do not want to *know* that Avril Lavigne is feeling frustrated because her boyfriend acts differently with her when his friends are present, we want to *feel* that frustration.

So how does this authenticity of the artist come through? It is all too easy through the ideology of rock to say that the crossing of genres that The Matrix did with Korn lacks any sense of authenticity, that it speaks only to following success, to following the money. It is unlikely that most rock fans would be comfortable accepting an artist that had turned from pop to metal, or at least that is not a commonly accepted transformation. For example, the Dave Grohl school of thought might say that musicians are best served not by "standing in a line for eight fucking hours at a convention centre" for a talent show only to be told that you're "not fucking good enough" (Grohl 2013), but rather that success is achieved by working really hard in your garage with your band. This is rock ideology in action, but of course Grohl is not a tyrant averse to pop or indeed a sense of humour, having invited Rick Astley on stage with the Foo Fighters to 'troll' his audience with his own live 'Rick-rolling' meme at the Summer Sonic Festival in Japan in 2017, with Grohl describing Astley as "his best new friend" (Billboard 2017). And more recently the Foo's 'Dee Gees' disco admiration filled tribute '*Hail Satin*' (2021) and even a comedy horror movie *Studio 666* (2022). If rock has always enjoyed being a parody of itself, or at least for the most part not taking itself too seriously, then its wide spectrum of fans also includes some humourless purists to whom betrayal is a possibility ('Judas', one might shout). So, what is 'authentic' about a production team, or indeed a rock star, that is not *faithful* to a particular stylistic identity? The opportunity to explore this came in the form of the nu-metal band Korn who asked The Matrix to produce their seventh album *See You on the Other Side* (Korn 2005). This was not a unanimous decision by Korn band members, in keeping with their rock ideology there was a clash of authenticities but lead singer Jonathan Davis persuaded the band to go with it, as the *New York Post* reported:

> When I told everyone I wanted to work with them, everybody was freakin' out, [...] They were like, 'Are you doing pop?' and I'm like, 'Shut the [fuck] up and do it.' Music is music, it doesn't matter what the [fucking] type it is, so let's see what happens.'
>
> (Shen 2005)

Style then is not privileged by The Matrix but rather it is the capturing in music of emotion, aggression, and passion. Rock has as a central tenet Allan Moore's 'first person' expression that the artist is "telling the truth of their situation" (Moore 2002), but that is not an ideology held exclusively by rock. Through Moore's types we might consider The Matrix to be nearer to that of third person authenticity in that they believe in the passion and expression of *others*. That is, for the most part, their role is that of making others be as genuine as possible. The Matrix have at their core their passion for music and with that comes the understanding of other people's passion for music. As Mia says to her soulmate and jazz musician Sebastian in the film musical *La La Land*, "People love what other people are passionate about." (Chazelle 2016). The context being that success is not about chasing the audience, chasing popularity, but rather that the audience wants passion, often regardless of how that passion is made manifest. And so it seems the skill and belief of The Matrix is to draw out that passion and authenticity from their artists. They believe in passion for music not in any one particular style. For them to "capture emotion and aggression" they first need to find its authentic voice, and the most documented of their successes, that of Avril Lavigne's first album *Let Go* (2002) hits 'Complicated' and 'Sk8ter Boi', demonstrate why their approach worked. As Christy recalled:

> We'd been listening to the kind of stuff she had been doing – it had a Faith Hill kind of vibe, and as soon as she walked in the door we knew this was just wrong. This kid had melted toothbrushes up her arm, her hair was in braids and she wore black skater boots. She didn't seem like the Faith Hill type.
>
> (Buskin 2006, p.142)

The recognition that Lavigne's style at that point was not her authentic self put The Matrix not in the role of 'style finder' but rather as that of 'authenticity facilitator'. There were no guarantees with Lavigne of success and up to this point the team had not had major chart success, but the questions they displayed in their approach turned out to work, that is, to help Lavigne find her voice rather than to give her a voice based on perceptions of audience demand. When we speak of 'Internal coproduction' we note the presence of the past and the future in our decision-making that relates to others, to an inauthentic '*they*-self' in Heideggerian terms. Here the emphasis is not on the future listener of Lavigne's recording, or indeed past knowledge of the listener, but rather the collaboration is with the artist and *their* internal voices (which of course may very well include past and future audiences). It may only be a hypothetical discussion to wonder what the team would have done if the voice that Lavigne had found was not one that resonated with the public and that her preferred style of expression had *not* been as likeable and the record *not* been a success. Would that have changed their approach, or would they have stuck to their core belief, upholding authenticity and passion regardless of potential commercial failure or would they have changed tack to 'please to punter'? Possibly out of our own bias when writing about those one admires, it seems that even when Christy is discussing writing 'hits' that it is not the popularity that is the goal but the wanting to do one's very best at writing that one can. In an article in the magazine *Miroir*, Christy said that "I always try to write a hit, you know, I'm not interested in writing a song that could just be a nice album track" (Websman 2013) but this is preceded by her words that "Every day I come to work and try to give it all I've got", it underpins that the goal is not about the success of a hit but about the wanting to write the *best* song, of which being a hit is but one measurement of success. The *Miroir* article also notes that "This year [2013] alone she has written more than 150 songs". Her songwriting can be prolific. Taking this speed of production back to working with Shakira, Christy notes that they wrote seven songs in one week with the artist. An intense pace, which puts at risk the need for songs to be about something relevant to life. As she pointed out:

> We don't work like that all the time. You need the down time to live life, to be able to write about actual life experiences. It would be boring to write about just going to the recording studio! [...] We spend half the time just writing new songs on our own which we save until we pull out a song when it fits for an artist. The other half of time, we spend co-writing songs from scratch with the artist. When it comes to our writing together, there are no real rules.
>
> (Kawashima 2006)

The rules are that 'there are no rules', though of course this generally translates into the ability to flex according to the situation. Spock emphasises that it is not rules that drive The Matrix but an adherence to their belief, anything they do must meet agree with their headline statement "Our passion is music", if it falls outside of that then they don't do it:

> The common thread running through every project we do is emotion and aggression [...] Both are present, even in a ballad. But we make music that is also very accessible. We take the things that are really cool about an artist and focus on that, distill it down to its essence, without ever taking the artist out of the music.
>
> (Yamaha.com 2018)

It is not only their passion but that of the artist that they believe in (the passion of others), regardless of whether that is expressed through pop or rock.

There is, as it appears at first, an anomaly in this 'authenticity' approach to the artist. In 2004 they began work with singers Katy Perry and Adam Longlands on an album, but the project was pulled weeks before it was due to be released in September of that year. In *Katy Perry: Purposeful Pop Icon* (2019), Valerie Oswald writes:

> Perry was disappointed – but also relieved. The style of The Matrix did not really fit her own personal tastes [saying] 'I had this kind of quirky, unique perspective, and they had a very mainstream pop perspective, which was really cool, too, but I wasn't used to it [...] My own stuff is very heart-on-my-sleeve'.
>
> (Oswald 2019, p.24)

It suggests, in contrast to their stated ethos, that The Matrix were 'imposing' a style rather than facilitating Perry's authentic voice. But the actuality of the situation is a clash of authenticities not a lack of one. The album was self-titled 'The Matrix', hence the song-writing and production team were the band, the artists, so what was required of this project was 'their' voice not Perry's, and hence the clash of authenticities that led to the shelving of the project. At the time with the album due for release, in an interview with Perry and Christy for *Ink19* by interviewer 'Sir Millard Mulch', Christy was asked about the reasons for writing their own album, to which she responded "Rather than always trying to get across another artist's point of view: 16 year old, 24 year old, guy or girl, whatever. It was just interesting for the 3 of us to go, 'What do WE actually want to say?'" and then with regard to whether there would be a tour of the album Christy responded that:

> We have to be honest about what we are, and what The Matrix is – people who have written hit songs for other people. And to suddenly just come out as a grass-roots band that is playing little clubs everywhere and trying to be cool would be totally FAKE.
>
> (Mulch 2004)

Even though all three of them had successful professional careers as artists before becoming a production team, there is still a sense of the need to be 'true', a need to ensure that the audience is not fooled, that simulation is not at play. So, whereas they had a transformative vision for others, in this instance the project was to apply that aim to themselves, and that authenticity is theirs and not another's. This proved to be the barrier at the time and the album was put to one side, even though great expense, including videos of the songs, had been made. The Matrix decided to shelve the project.

> We'd all worked really, really hard to get to this place behind the scenes, and our own project just took way too much time and actually kind of affected our career for a little bit. We therefore decided to pull the plug on it and get back to doing what we love to do.
>
> (Buskin 2006)

The album was released five years later on their 'Let's Hear it' label and remains something of a hidden gem of pop music of that time. In an interview for the BMI's 'How I wrote that song' series in 2017, the pertinent question was asked with regard to how does the songwriter 'authentically capture the voice of the artists?' Christy puts it down, in part, to recognising something in the artist's ideas that they, the artist, might not have recognised themselves:

> I do a lot of listening, probably more listening than I do talking in a session [with the artist], and I'm really good at spotting other people's genius, when they come up with something that's brilliant and they've gone past it, I'm like "Back that up, on your voice recorder because I just heard something amazing" so I'm really good [at spotting those moments], I call it being a butterfly collector.
>
> (BMI 2017)

As part of the team that Christy described as 'the three-headed monster' they seem to have managed to catch the butterfly, found its voice and projected its passion, emotion, and aggression. We might have to rethink our cultural perceptions of what a butterfly is, not as a fragile thing of beauty to be protected by the benign generosity of those with the power to do so, but as a collaborator with monsters capable of telling it like it is, and making us feel it too.

6 Hierarchical Production and Complementarity, Before, During, and After PWL

An Interview with Phil Harding

Christopher Johnson and Robert Wilsmore

Introduction

Phil Harding (now Dr Phil Harding) is a mix engineer and producer whose collaborative experiences have covered a wide range of types from the distributed, leaderless, and chaotic approach of punk to the highly structured and hierarchical factory model of the hit machine Stock Aitken Waterman. Harding was there at the very start of the Hit Factory and the PWL (Pete Waterman Ltd) studios that dominated the UK pop scene in the '80s, having been drawn across from his position as mix engineer at the Marquee Studios in 1985 by Waterman, where he had worked with bands such as Dead Or Alive that were to become inextricably linked to the SAW machine. Whilst at the Marquee studios he also worked with one of Pete Waterman's idols, Lamont Dozier, from Motown's Holland Dozier Holland collaboration, with whom Phil would work many years later mixing for Lamont Dozier and Sir Cliff Richard on the 2012 *Soulicious* album. Before the pop days of PWL and his subsequent work at the Strongroom with Ian Curnow, manager Tom Watkins and, in particular, the band East 17, Harding had other vastly differing experiences with punk bands such as The Clash and Killing Joke, and then later, after punk and post boy band, on his own projects where he crossed over into performing and songwriting. The collaborative practices and differing structures of these experiences are detailed in the interview and commentary that follow here.

In all the literature and anecdotes that we have read we find many stories of the fiery nature of those he has worked with, the egos that demanded this or that, and that clashed in the heat of the moment. If anything comes across about Phil Harding, it is that his passion works in tandem with his professionalism, and we are left wondering if his calming influence, good humour, and diplomacy in those fraught moments might be somewhat underestimated in the positive effect it must have had on the production process. An assistant engineer at PWL, Les Sharma wrote of 'the mix master':

> Phil Harding is the nicest man I have ever worked with in the music business. It was an honour to work with him for 3 years. He took time to explain how records were recorded and mixed and was the best teacher you could want if you were an assistant engineer. I can say he never lost his temper in the 3 years I worked with him, he is a kind, honest and gentle family man.
>
> (Sharma 2014)

More recently Phil has stepped into the academic arena, undertaking his PhD and working through his experiences with a diligent, pragmatic and detailed approach. His account of PWL in his book *PWL From the Factory Floor* (Harding 2010) is quite different to the more typical autobiographical publications of, for example, Pete Waterman's *i wish i was me: the*

DOI: 10.4324/9781351111959-7

autobiography (2000) and Mike Stock's '*The Hit Factory: The Stock Aitken Waterman Story*' (Stock 2004). Harding's writings on the other hand, as well as covering the interesting aspects of the tempers and the inspirations and the in-fighting that others cover, also includes much of the technical detail of the events as well as deep insightful analysis. His writing seems to echo the man and his contribution to the productions he worked on: rigorous, detailed, comprehensive, observant, honest, approachable, good humoured, passionate, and professional.

We met Phil at the Headingley campus of Leeds Beckett University on 8 July 2018 where he had been leading an event as Chair of JAMES (Joint Audio Media Education Support). Our interview did not have a formal start, rather we segued from a discussion about interviewing into a discussion about collaboration. Phil had himself recently been interviewing producers and managers for a book based on his PhD, now published in this series of Perspectives on Music Production (see Harding 2020), and hit upon the problem of transcribing interviews when much of what is being communicated is not spoken, so it does not appear in the text. And that, more or less, gets right to the point of our study; how do collaborations operate when so much is unspoken? He laughs (kindly and endearingly) at the world he has stepped into whilst doing his PhD.

PH: Much as I find with most academics [laughs], no disrespect, they have a tendency to over complicate things. But that's fine, it's the world I have stepped into. [...] The problem with transcribing interviews is that the language, especially when two people have worked together or are on a similar plane, there are tonnes of sentences that don't get finished, so for it to make sense the interviewer has to fill in those spaces for the data to be accurate.

On the first point he is perhaps right, we are as academics trying to be clever; however, we should defend our position by saying that we are striving for better understanding, for new insights and knowledge. That we seem to overcomplicate things and appear to be 'trying to be clever' is a result of pushing our own abilities as far as they can go and appearing somewhat clumsy and not quite as clever as we would like to be in the process, but our aim is not to be clever for clever's sake. On the second point, our approach has not been to transcribe the whole (2 hour) interview here but rather to transcribe pertinent parts, editing only for concision and to represent the interview in words as well as we can in order for "the data to be accurate". Neither has it been left as a transcription but we have chosen to add commentary throughout in our search to understand the collaborative practices being discussed. For clarity, as we move quite freely between interview and commentary, the interview transcription is given in italics with commentary being non-italicised.

'Telepathy'

Given that the informal discussion that our conversation started with led us quickly to non-verbal communication, it seemed appropriate to segue straight into one of our questions about how long-term collaborators develop an understanding that means verbal communication is minimal or reduced to nothing. We noted in the first chapter of this book that he and his collaborator Ian Curnow had a working relationship that was almost 'telepathic', and Pete Waterman also notes this, writing of himself, Mike Stock, and Matt Aitken that "from the beginning we began to understand that when we were on form all three of us could work together almost telepathically" (Waterman 2000, p.138). We asked if, assuming that 'telepathy' was an analogy rather than a fact, what was meant by this.

PH: That really only comes with, let's bring it down to two people, both being experienced, where they have been working and collaborating with each other for some time, and that's not going to be instant. So, if your producer or production team is with a new artist you've got to break those barriers down, but within a production and writing team it's quite incredible how little needs to be said, and the conversations tend to be based around the problem area rather than what needs to be done. In other words, if one party is expecting the other party to do a specific thing and it's all sounding great or it's quickly editable there is no discussion needed, you're just getting on with your job.

Within this factory model, part of this telepathy is simply that of replicating previous behaviour, where it has worked in the past then that can be replicated without need for validating interactions or problem-solving between collaborators, each contributor is 'getting on with their job', and as that has prior approval (however tacit) it does not need a new approval each time, hence the team can work at speed. It is only when something falls outside of that (in a good way or a bad way) that the contributors need to interact, problem solve, and reach an agreement (or have team leader approval).

Team Leaders

Having a strong team leader then increases productivity in this respect, the collaborative process is not equal, so less time is spent on negotiating agreement, an executive takes the decision and the project moves on.

PH: In pop it's really based around a team leader and they are the one that is usually gelling the collaboration together, no matter how many people that is, and therefore is the person feeding the instructions to the creative team. That team leader is feeding the project [...]and expecting a certain result, not instantly but the next time they walk into the room, if it's all there then there is nothing to be said, if there are certain things expected and they are not there then the conversation needs to happen. And if things that were not expected, but the team leader thinks were good, then that also caused a discussion. So, there were negative and positive points that led to discussion.

I have come up with this notion of a 'service model' based around production collaboration, and in that particular model [PWL] there was a very clear team leader and that was Pete Waterman. A shining example of an entrepreneur who doesn't want to spend all day in the studio, they want to be walking in and out feeding their instructions and visions for the project. Because that's the person walking in with the cultural capital, going to the Pierre Bourdieu references, the *social capital, and* probably the economic capital. Certainly, in this case he [PW] owned the studio, he owned all the equipment. Obviously, everyone was working for him, even his partners Matt Aitken and Mike Stock, and that's why there has been so much friction since, because that was a partnership based on a pretty loose business agreement that ended up being argued in court. But funnily enough, they keep coming back together.

Pete Waterman described himself as a "dipper" (Waterman 2000) with regards to his presence in the studio, and here the skills of Waterman are complementary to the others. He uses his social and cultural capital for the advantage of the team, none of whom (at this point) are in a position to take this role, neither having the time, the connections, or the status (social capital) to do so. Recognising the value of this role was key later to Harding and Curnow appointing Tom Watkins as their manager after leaving the PWL set-up. Business and creativity are not two completely separate aspects of a project that are disconnected through division of labour, but rather they are bound up synergistically. It might even be said (such as Warhol expressed) that the art is the business and the business is the art. Certainly the art of hit making contains both creativity and business as predicates of its being. We can see division, for example where Waterman is dealing directly with the client with regards to money (backed by business manager David Howells), and Stock and Aitken are working on the creation of the product, but there is also a unity, a 'common vision and trust' as John-Steiner describes in her Family model of collaboration (2000, p.179) or it might even have been the Integrated model of 'visionary commitment' between Waterman, Aitken, and Stock. Certainly, Waterman was committed to his vision, how the rest of the production team might be described probably varied from being 'committed' to it to 'sharing' it.

PH: Talking about hierarchy at PWL, underneath were Mike and Matt creatively and David Howells who was the managing director really holding the business end together behind Pete. I don't think you can ignore how a creative team is working together business wise because it is going to affect the work, how they work and why and when they work. [...] No one was allowed, or meant to, bring their own equipment in. Pete would also say to myself and Ian Curnow "If you want something, give me the rationale for why we need it, why we want it, if we need something to be competitive I'll go and buy it but don't bring your own in". His concept as a business person was 'I'm renting my studio with my equipment, to my clients, which my teams are fulfilling'. Pete wanted ownership of everything. So then we moved on to the '90s where we 'owned' a studio, we had a room at the Strongroom, it was our equipment and we were in control of our own destiny but we did decide to work with Tom Watkins because we knew pretty quickly after PWL, where Pete Waterman was managing us, that we would need a manager. At a certain level of production, you are spending all your time in the studio, you haven't got time to be dealing, certainly as a two-person team, with the business and the clients and judging what are the good gigs and what aren't. That's another role that someone like Pete Waterman takes. With a successful production team you've only got so much time. So next week if you've got an option of four different projects, which ones have the biggest marketing spend, commitment, budget, record company, etc. behind them, because those things become the choices as well as the creativity. [Tom Watkins] is another entrepreneur as a team leader with a different mindset to Pete Waterman, he wasn't interested in owning a studio, wasn't interested in owning equipment, because he was an artist manager first. [...] We had a three-year run which was fantastic, now I label him in my service model as a 'team leader', but in a more aggressive way than Pete.

One of the characteristics of this team leader is the dominant and possibly aggressive way with which they ensure that they are in control. As Waterman described it, he was looking after his vision whilst others might have seen this as 'pushy', and he writes "I don't lie to myself, I don't lie to the people I work with and I don't lie to the artists. So the people I work with here might call me a bastard to work with [...] it's that attitude, that bull-headedness, that gets me disliked" (Waterman 2000, p.203). In truth he does not seem to be disliked at all, perhaps it was understood that the bluntness was not personal but was for a greater purpose (the 'vision'). That characteristic in a team clearly has advantages and Harding and Curnow went on to appoint a similarly strong character, Tom Watkins, in their move away from PWL and SAW when they were partners at the studio in the Strongroom. Phil describes interactions with Watkins in this manner:

PH: I took some very heavy phone calls off him [TW]. Your first priority, day to day in the studio, is keeping the artist happy, your end goal is keeping the client happy, or if you're not dealing directly with the client your mission [in the cases of PW and TW] is to keep the team leader happy. Then the team leader goes off and makes sure the client is happy and they are bridging that gap between the client with the money and the final decisions. So, for you as the producers, in the studio of East 17 for instance, you are hell bound, on the vocal days, to be keeping the artists happy, getting good vocals out of them. So, there we are pandering to all the artists requirements and needs and all the rest, so Tom Watkins set up an office just round the corner to us in Shoreditch (and not a small office). Once East 17 were on to their second album he went from an office with two assistants in Maida Vale to an office with 10 or 12 people working there. So, what would happen, for instance, East 17 would finish recording and wander round to the office and start getting into conversation with Tom about 'Phil and Ian were saying this …' and then I get this phone call that I had been leading his artists astray. Nothing to do with the music or what's going on creatively, and we'd quite often get an earful on those occasions.

Speed of Projects

The strength of passions in these production teams was in part due to the business nature of the productions. They were hit 'factories' and like any factory there was a direct link between time, money, creativity, and productivity. All of these factors were important. The speed at which SAW operated is often documented, in particular the Dead Or Alive recording of 'You Spin Me Right Round' (1985). As such we asked whether the time frame was something that was deliberately imposed in order to achieve an urgency and hence a dynamic product. That does not seem to be the case, rather Harding gives an anecdote that expresses how they might have increased productivity but through delegation and expansion rather than squeezing a time frame. His example is post PWL whilst at the Strongroom.

PH: Tom Watkins said 'if you guys would just take on more people I could get you two or three times the amount of work' and it's easy to say now, and we probably didn't handle that as well as we should have done, but we probably should have taken that advice on. But I think that amongst those decisions was Ian wanting to keep a very tight handle on the technology and the programming and he had a particular way of working that we would drive into our assistant, but anytime we got someone else to do a bit of programming Ian came back to it he'd say "oh, I've now got to spend hours reworking it" because they'd not done it the way that Ian had set up. So myself and the assistant would know exactly what we should be doing with Cubase, someone else would come in and work in a different way and actually that's taking more time than saving time. […] but never once in the three years of collaborating with him [TW] did he come in and say "you're too slow". If there were any tight schedules I don't think we ever missed any.

It is a typical business conundrum; how much can one increase productivity by spending time training others compared to what one can achieve by doing it one's self and avoiding spending time on training. Harding's view here suggests that maybe the training option might have been beneficial in the long run. We asked if the speed of production at PWL continued naturally into the Strongroom productions, rather than as a deliberate strategy?

PH: Not really, we [PH and IC] didn't have those ambitions whereas Waterman did. Initially in the first year or two [at PWL] I was unofficially chief engineer but doing most of the mixes, and we'd have recording engineers doing the recording. But once myself and Ian got into more remixing and production it was clear and obvious that we'd got to get more mix engineers in. That's when we got Pete Hammond in and then Waterman persuaded Pete Hammond "Oh, it would be so much more efficient if you worked overnight" [laughs], and the same with Dave Ford about a year later. So suddenly there was a choice of three experienced mix engineers, and you would see from ''87 to ''89 and a bit onwards, Dave Ford and Pete Hammond doing a lot more of the direct PWL mixing than myself.

Rather than individuals, the expansion of staff that could fill certain roles and the product that was still identifiable as a SAW product led us to ask whether elements within this structure could be replaced, could a mix engineer or an artist be substituted and the brand identity remain intact?

PH: A number of the major mixes might have gone through more than one mix engineer's hands [...] My memory was that, for instance with Mel & Kim's 'Respectable', when you're doing the follow-up single to a first debut that's been a big hit, there's this extra pressure that comes from the company, from the PR people, that hits the whole team, of expectation. It's got to be at least as good if not better than the first one. [...] And with 'Respectable', I think Pete Hammond would agree with me, certainly he and I mixed that a number of times and the difference being that he would claim that he would always start again from scratch but my view was, that if the mix has got to a certain level, whoever's done it, and Pete's [PW] close to happy with that but some adjustments are needed, then I'd say to the assistant we'll recall Pete Hammond's mix and we'll work from there. But when Pete Hammond's been asked that question "would the assistant recall a Phil Harding mix?" he claims 'no'. I can't answer for him but certainly 'Respectable' went through each of our hands at least a few times. I think we ended up with a co-credit but it's debatable as to who did the final one.

The truth as to who did what, if there is one, is literally lost in the mix. The process of mixing and mixing again either from scratch or from a previous mix throws into light a collaborative process whereby contributions are individual in terms of time, that is, they worked sequentially passing on the mix rather than working on it at the same time. So, had the process been documented we might indeed be able to follow the series of events that led to the final mix, but that was never a consideration and the contributors merely 'got on with the job'. And without documentation, without 'evidence', the dominant account is that both parties worked on it, and hence a co-credit is a reasonable acknowledgement. Perhaps the most complex of integrated collaborations, such as John-Steiner describes between Braque and Picasso, could be similarly split. What is not captured is how the previous decision of one affects the decision of another. If the final version was a Pete Hammond 'strip it down and start again' mix then influence might not be a part of the product, though of course it might well be that influence is still at play by learning from what was before and not repeating what one did not like about it. As much as anything this particular example outlines the fact that coproduction may well be an attribution based on the fact that no one really knows who did what.

The 'No Man's Land' of the Middle Eight

So, even though the PWL and SAW model had complementary roles, they were obviously not *always* clearly defined roles (or at least there was some implicit and acceptable bleeding across roles). In the previous example both Harding and Hammond had the same role, that of mixing engineer, and the confusion falls within that category (it was all mixing, it's just not clear who mixed what). We asked if Phil could explore this further – were there roles that crossed over into other roles?

PH: If you listen to a bunch of PWL records there's no traditional American-style middle 8. It became something that was, in my view, a bit of a lazy factor from Stock and Aitken. At times Waterman just let it go, but within the song arrangement there's no traditional American style middle 8 with some different movements and different chords with a set of lyrics and different melody. You find me a traditional American style middle 8 in a Stock Aitken Waterman song, if you do it's a rarity. So most times what would happen is Mike and Matt, knowing that people like myself and Pete Hammond are coming in and mixing, it almost became a running joke or you could almost call it a competition, they'd expect us to fill 8 bars in-between the second chorus and the last choruses, which generally would go back to the verse movement, might carry on with the chorus but probably back to the verse movement and probably come out of that vocally back into the bridge and then to the last choruses. [The 8 bars were left] vocally and melodically empty and they would expect the mix engineers to sample something from elsewhere in the song and to fill that space. So, listen to two or three of them and you'll hear triggered vocals, backing vocals that have been sampled from elsewhere. 'Never Gonna Give You Up' is a classic example [PH sings the example "never gonna give, never gonna give"].

These then were tasks that could be traditionally grouped under 'arrangement'. So we asked if there was an understanding that the mix engineer does that bit of the 'arrangement'?

PH: Absolutely, which takes a bit of pressure off Mike Stock and Matt Aitken's shoulders. "OK, we only have to take this so far, and those guys will finish it off and we'll make a judgement of it once we hear it" and if someone's not happy they might have to step in and do something themselves but that decision would generally be up to Waterman. [...] It was an unspoken "There's a gap, fill it. [pause] Same as you did last time!" [laughs]. Which obviously is a new creative little job for the mix engineer, you're not just hunched over the desk doing all the technical stuff. So, amongst the first things you'd do in that situation is, you might say to the assistant "let's get sampling some bits from the chorus or some other lines". And one of the things that made that easier, when that continually kept happening was the Publison Infernal Machine, unlike the AMS sampler, you could put in a 10-second sample with a manual trigger and literally one of you is on there doing it [PH mimes pressing the trigger button].

The example highlights the nature of production as performance as well as of composition. The mix engineer has both compositional input and performance input, but essentially these tasks are merely delegated by an executive who has the official title of producer and, of course, the credit and the royalties.

Collaborative Models

The complementary nature of the collaborative model is clearly defined in the organisational structure and then it has acceptable limits of crossover (implicitly agreed) within that. We asked if the roles changed over time, as might be expected with collaborative groups that go through the typical Tuckman stages of forming, storming, norming, and performing (see Tuckman 1965).

PH: The early days were different to when it moved on. [One keyboard player] no longer wanted to come in because there were some fractious times, part of the SAW model was that assistants and everybody had to go out of the room so that the three of them could thrash it out. We never knew what happened behind those closed doors. So, he got to the point where he found that difficult. I'm saying that because both Mike Stock and Matt Aitken were both very able keyboard players, but I think there was an understanding, certainly in those early days, that, OK so they were great keyboard players, but did they have the soul groove that someone like Andy Stannett who had been in Freeze [PH sings 'IOU'], and he was great, he played bass on the Rick Astley stuff and keyboards on lots of Princess stuff, Mike and Matt knew all the notes, knew all the chords, but didn't necessarily have the groove. After Andy went a range of keyboard players would come in to take the pressure off the guys. Because you've got to bear in mind, especially if it's a song they've written, everything from the first note onwards is driven by them.

The collaborative process I saw for the style of a SAW song was getting into the mind of the artist, the character of the artist, and that would be a discussion with the three of them [Stock Aitken Waterman], and quite often Mike looking to Pete to say "well, feed us a subject, feed us a line, feed us a title, feed us something we can hang a chorus on", although that began to happen less and less after '87 let's say but certainly up to that point, 'Never Gonna Give You Up', these sort of phrases, would come from Waterman and then having decided, this is the character of the artist, this is what we're heading for, he would walk out of the door and leave them to it.

The hierarchical structures and the roles are clearly delineated here; the clarity of who was in charge of what, who has the responsibility (and was given the credit) and hence where the executive decisions within those parts lie.

The Persona and the Pop Machine

Our probing around the interchangeability of roles (engineer, artists, etc.) comes from one of the eternal critiques that hangs over pop music, that is, that it is not driven internally by the artist but externally by market forces. Phil goes over one of the classic examples in SAW history for us to show how the persona, the singers, were central to the method. It might be argued that this is similarly inauthentic, that the singer is still interchangeable but that the deal is that, whoever they are, their character, life experience, etc. will be drawn upon as part of the material. How this is drawn out though is of interest in a collaborative study. In this example the method of arriving at the lyric 'showing out' in the song 'Showing Out (Get Fresh at the Weekend)' comes in an informal moment, a moment of 'distributed collaboration' in John-Steiner terms.

PH: Mel and Kim are a fantastic example of them writing to the character of the artist walking into the room. They [Mike Stock and Matt Aitken] had written a song called 'System' but hadn't really spent a lot of time with the girls before that. [At that time] Pete Tong had brought in the first Chicago House records to Waterman and he said "let's do this with Mel and Kim. We're gonna write a new song based around Chicago House and we're going to get the character of the girls". The actual term 'showing out', and most people agree with this, came from the assistant Jamie. One of the girls said to the assistant "What are you doing at the weekend Jamie?", and he said, "showing out". Mike Stock heard that, and it goes from there. No one in the room other than the girls understood what Jamie said, what he meant was he was getting dressed up, going out clubbing, showing off, the rest of the guys in the rooms weren't doing that at the weekend but the girls would have been.

RW: This was an informal, voluntary contribution, was that typical?

PH: I can pick out occasions like that because they are well known but generally it's ground out. [...] All the songs were written for the character of the artist and there would be a planning discussion about that between themselves, between Mike, Matt and Pete, having met the artist. The more that the artist came in and the more Matt and Mike got to know them, the more they could write stuff that they felt suited the artist, their character, and even what was going on in their lives. And they're not really given credit for that.

PH: Those conversations [between Stock Aitken Waterman], apart from at the start of the day, would happen at the traditional stopping point at 10 o'clock, and unless there's some type of emergency going on, it's all down the pub. The three of them would sit round and chew over what's happened for the day and talk further about more songs for the artist and so on. Although a lot of us would go to the pub it would often be the three of them and everyone would leave them alone.

Upstairs Downstairs at PWL

The separation of the songwriting team SAW and the others at PWL literally has a sense of upstairs downstairs to it. Just as 'the grown ups' lived up the stairs in Abbey Road Studio 2 control room and the bands performed downstairs, so the set-up at PWL has Stock Aitken Waterman upstairs and Harding and Curnow and others downstairs in the Bunker studio and other rooms in the basement (see Harding 2010, Section D 'inside PWL studios' for exquisite details). Harding's collaborator Ian Curnow recalls those moments where "They would come down to the mix play-through with their arms folded" (Harding 2010, p.468) and their place down the pecking order became apparent to all with regards to the control that Stock Aitken and Waterman had. A position further highlighted by the fact that Waterman's office from 1986 to 1989 is described as "PWs office, with throne" (2010, p.416). Phil notes how the method became more separated as time went on.

PH: In terms of the SAW songs we [Harding and Curnow] would just be pulled in for the latter end of it. It was only in the early days when we only had one studio and before Ian joined that I would pick up any information about how the songs were forming at the beginning. [...] Even in those early days, '85 and '86, we would get frustrated with the communication from Pete sometimes because he can't go to a keyboard, he can't do much technically on the desk. We were coming out of one project at the beginning, there were some frustrating projects at the beginning, certainly with a band called Spelt Like This, managed by Tom Watkins. But after a brilliant project with Jimmy Cauty, something that happened that Mike and Matt and myself were involved in, one of the backing vocalists Princess got together with Pete and he decided to make a track with her. It was pretty much a blank week and Pete was going on holiday, and there was this sort of relish between Mike and Matt, so they got the instructions off Pete, what we should be doing, where we should be going, what tempo, and it was a soul project so it was a very obvious Jimmy Jam and Terry Lewis style. So, we had a number of records [to listen to] that he had left behind and he left the team to it for a week. It was a breath of fresh air; we had a whole week without Pete, but we wanted to do something really good so that he was really impressed when he came back. So that first playback to Pete of 'Say I'm Your Number One' was one of those big moments that sticks in your mind because Mike and Matt knew they had largely achieved it without him being around without all the daily collaboration. [...] What we liked was that without the daily visits form Pete we got a clear open hand but with the song started and the references all set. But that was early days and that was pretty rare.

As time went on, '88 to '89, to the Kylie and the Jason [period] there would be less and less of Pete coming in and out, there would be more of the factory 'go through this process, go through that process'. A finely tuned product could be rolled out. You asked earlier if the artist could be interchanged and people commented a lot about that. It really only happened once or twice, the famous instance being when Rick [Astley] stormed out and it was given to Jason [Donovan]. [...] '87 was the peak 'flow' in terms of creativity. We may not have hit the peak in terms of economics but it was the peak of creativity, we were riding high with Rick, Mel & Kim, and one of the key things we are remembered for is Kylie, but when Kylie came along we lost the dance community, our start point in the mixing process was the 12 inch. Almost everything up to Kylie was a 12 inch first then the 7 inch radio version cut from that.

Integrated Model: PWL as 'The Band'

We read out a quote from Bennis and Biederman's *Organising Genius* (one that Vera John-Steiner also quotes in the discussion of her integrated model of collaborations), and asked if this resonated with his PWL experience:

> People who have been involved in Great Groups never forget them, although most groups do not last very long. Our suspicion is that such collaborations have a certain half-live, that, if only because of their intensity, they cannot be sustained indefinitely.
>
> (Bennis and Biederman 1997, p.216)

PH: Yes, in the studio it is not dissimilar to what happened to a band on the road, that is a typical story of the band, and it does happen collaboratively in the studio with teams because it becomes intense [...] whether someone's unhappy with a business situation or a creative situation or a personal situation, there are so many factors. That's why, in my view, whether you relate it to a production team or to a band you need a good team leader or a band leader. Bands that have outlasted the typical model, like U2 or more recently Coldplay, those bands, have been structured, I'm assuming with good advice from management, saying "this is the band leader, this is the singer, this is the lyricist" and probably, whoever they are cowriting with says "let's stay together as a team, co-credit the rest of the band and decide how to split it". [...] In answer to what you've pointed out that these collaborations, when they're really integrated and tight that they don't last, that one can say that about boy bands. Would East 17 have lasted longer than three albums if that had been in place [an agreement] from album one? I think so.

CJ: Were you, PWL, the Strongroom, the band?

PH: Yes, and Mike and Matt have stated that really we were the band. If you look at the modern day R&B, hip-hop model in America where the producer is the artist and the singer is 'featured'. But that determination and front from producers didn't exist then, but the whole Stock Aitken Waterman story could have been SAW featuring Kylie, SAW featuring Rick Astley, but that's the pop world and it's been that way since Motown. The singers turn up and sing. And Ian and I could say the same for all the artists, all the boy bands we featured in the '90s, it was just the 'vocal' day [that the singers contributed too].

In retrospect the model is an obvious one, the producers are the artists, the performers are the raw materials, the stone the sculpture sculpts, the paint the artist paints with, etc. But it has taken time for the recognition of the producer as the artist to take hold because the separation of the artist on one side of the glass and the producer on the other has set that division long into our societal understanding of what a recording is, roles structured by the doxa of the Production Habitus.

Equal Co-credits

But in the late '80s and into the '90s it was clear that attribution of contributions were *not* going to be acknowledged in a non-hierarchical way. At least not in the pop world as Phil describes when attempting that post-PWL with East 17 and Tom Watkins.

PH: We said to Tom at the time "if we want a great album why don't we say that every song, if it's the ego thing rather than the business thing, that every song is written by the seven of us, the four band members and three producers". It didn't go beyond the meeting with Tom, it was kicked into touch, wasn't even discussed.

And when the egos kick in, the credits are not divided as fairly as they might have been, and the intensity of the team begins to show in the cracks then Phil describes with the third of the East 17 albums how he and Curnow fell back into 'service model' mode, a complementary skills model rather than a familial or integrated one.

PH: [In] an integrated collaboration where people are disagreeing with the way that the work and the collaboration is going about, well, you have to stick with it and grit your teeth, and there's a classic example of a service model. Ian and I were unaware of it at the time when we knew what was going to happen, but we accepted that album and took it on, we knew that we were in service to Tom, to the band, dropping our opinions, dropping any kind of ego, and delivering hopefully an album that was worthy of release.

Credits and Acknowledgement

RW: Pete Waterman, on a compilation album 'The Hit factory. Pete Waterman's greatest hits, 2000' thanks "... all the artists, coproducers and co-writers, record companies – who at most times, don't get credited yet without them we cannot be creative". Does being credited feel different now to then?

PH: A question of where you draw the line and where your mind is and where your ego is. Do you want the credit but not necessarily the money? My view was, for instance, with Dead Or Alive, where I did a huge amount of work on behalf of the team, especially the second album, where I'd be left alone in the studio with the band, mixing, and all sorts of things still needing recording; we could do drum overdubs and stuff like that but still needed some more keyboards and things, I had to go up to Mike and Matt and say "guys come down, this is too empty, we need things doing", and I'm just the mixing engineer. Even before then, when we were still at the Marquee working with Dead Or Alive, I took Pete aside and said "look, I don't want to disrupt the collaboration and the credits, but I'm that embedded in the mixing what about giving me half a point of the production?" At that time I knew that even half a point of a production royalty of something big could be quite a lot of money. For me that would have been a big reward. I asked on album one and by album two I knew I wasn't going to get that. It was a flat

'No' from Pete.[...]. Obviously, as a team leader Pete had to keep everyone happy. When Ian joined there were some similar problems.

Ian said to me that Matt Aitken had gone down and fed Ian some idea that went beyond just programming, there was a set of chords as well [that Ian had written]. Only SAW were getting writing credits. I placated Ian saying there will be some sort of 'pay-back'. In this case it was "here's three tracks to produce guys" and getting royalties [on a Rick Astley Album]. It was almost unspoken, it was "trust me and I'll look after you" but not specifically 'this is for that'. What a great team leader, and obviously our loyalty was maintained.

[Similarly] Pete Hammond could have said "I could be co-author of that song because I created the middle", all these kind of subjects for producers and engineers where you cross the line of producing or engineering or songwriting. You could write a whole book on that [PH laughs, knowing that's part of what we are doing]. We have a model now within the MMF [Music Managers' Forum] where acts that are still developing or on very low budget projects are better managed these days. They are often asking for 20% of the publishing before stepping into the studio and often getting it. If this is established then you are going to give more to the project. Otherwise, as a producer or engineer you're quite often thinking, I want to say this or suggest this idea, but if it crosses over into writing, and we haven't come to an arrangement, most experienced producers will know that you're never going to get it after the event.

At PWL we used the term 'additional production', it got added into mixing because so often we'd be stripping everything down to the vocal, and we would be saying to David Howells who would be negotiating those deals, can we get a royalty on this? And he thought about it and said if we can come up with a credit that includes the word 'production' then I can make a claim for a royalty. That credit came about because it meant a royalty and a fee. But if you're stripping a whole production back to the vocal and recreating a whole backing track why shouldn't you be on a royalty? And that would only click in if it became the main radio or video version, so not for a club version.

What Does a Producer Do?

Our investigation is founded on the study of uncovering contributions from different producers to a single product (in the widest possible usages of the term 'single'), and we always return to the multiple tasks, skills, and approaches that make up the complex and ambiguous catch-all term of 'producer'. So, although our question 'What do producers do?' sounds inanely oversimplified, it serves to draw out a reductionism to complement the minutiae of detail that we have otherwise been trying to draw out from our questions. So, what does a producer do?

PH: [laughs] When you say it to members of the public who really don't know, the easiest thing to relate it to is [that] a producer is doing in a recording studio what a film director is doing on a film set, even though they are two different titles, and Waterman believed this firmly.

So, what does a producer do? They hold everything together, specifically in the pop world, you're often a team and you're that joint between, whether it's the management or the label being the client and paying for the project, you're in-between them and the artist. For me most producers wear two hats, many people would say many more than that, but there's the one keeping the artist happy and the creativity, and the one keeping the client, the record company or the management happy, that's the commerciality hat. So, to me as a pop producer creativity and commerciality can't be separated.

It's always been easier in manufactured pop, where the artist is coming in and singing and doesn't really have control over the record, it's slightly easier to control it as a production team in that genre because, until they become successful and experienced they are often just happy to come in and take direction. It's only by album three they think they're great, they think they are responsible for the early hits no matter what and they're going to come in and tell the producers what to do.

Which of course, is one of the reasons why these intense collaborations begin to unravel after a relatively short time. Ownership and self-accreditation where there is success sews the seeds of the end.

Working With a Band

Although it is the days at PWL and the Strongroom and that factory model that dominates much of our discussion (and indeed what Phil is asked about generally), he has a long history of working with bands, both pre- and post-PWL. In this section the conversation moves towards the priority of the purpose of the recording, which can be reduced down to a binary between the priority of the recording and the priority of the unity of the group. In essence, where a difficult decision needs to be made between making the best recording by replacing a group member or accepting a lesser product but maintaining the group membership. The labelling of a 'weak link in the chain' depends entirely on what one identifies as the chain. A weak link in a recording might well be the strong link that holds the band together.

CJ: What's in your head when you are blending personalities into one entity?

PH: I have worked with a few bands in recent years and I almost prefer to step back and take the engineering role. There is a band [I've worked] with, they've all come in with a certain character and they are trusting me to record it and mix it but I'm not directing them as a producer. I'm adding ideas when I'm asked rather than pushing, and that's a difference between an engineering role and a production role, a producer can go out and ask and push whereas an engineer ought to wait to be asked, that's an etiquette thing. [...] We have got to the point now that the band are going to leave you to get on with a mix. And, especially if you've recorded it, you are thinking of the character of the musicians and how you're going to get that through in the mix. But it does cause some difficult problems if you felt that one player is not as musically efficient as the rest of the people on the record.

CJ: Do you have a strategy for dealing with that?

PH: That's a tricky one, your natural tendency as an engineer is to mix that down a bit because if I expose too much of this then the overall product is potentially going to suffer. [In one case] we got to the mastering room [...] and I know in my mind I've mixed [one part] down a bit, but it was immediately obvious to myself and the mastering engineer that we're lacking something. And I'm watching the mastering engineer furiously trying to get some bottom end and of course it was affecting the kick as well, and I had to step forward and say, "we've got a problem here".

CJ: Were you avoiding correctional editing?

PH: In an ideal world I would remove that character from the record and from the band and put someone else there.

CJ: Does the end product justify the means?

PH: Well, there's the difference, in this project I am only engineering. If I was producing it, I would undoubtedly replace the player because the end product is more important than anything, no matter what it takes to get there. But unless they ask me to produce I'm not going to present [any changes to them], whilst they're happy with the musicians they have chosen, it's their record.

CJ: I suppose [in this case] it is more that the band is having a kind of family collaboration where the most important thing is the continuation of the collaboration rather than the sales or the product.

PH: And that's what I'm saying about the two hats that a producer has to wear. How far does a producer go to ignore the group collaboration and the teamwork and the family, to do what they think needs doing to please the commerciality? It's a fine line.

Killing Joke: A Distributed Model of Collaboration

Knowing Phil's range of involvement in the '80s we should have expected to find the exact opposite of the reason we were so keen to interview him, which was to delve into coproduction within the factory model. But of course, one often finds the thing one was looking for when one is not looking for it. In this case it is the distributed model of collaboration, that is, a model that is informal, voluntary, with interested parties throwing in ideas and generally no centralised control, and certainly no obvious hierarchy of the Waterman model. But of course, before that era Phil was involved with engineering for punk bands including The Clash and Killing Joke. The '70s had given us binaries of extremes, the punk protest against the prog rock institution as well as disco kitsch and unison dance moves.

PH: Going back to the first Killing Joke album in 1980, [there was] no producer in the room, much like the punk attitude of just book us into a studio with a young engineer, we'll produce it ourselves, and the record companies didn't stop it happening. The Clash were amongst those that did that, I worked with them and there was no producer. And of course, as young engineers you've been trained, certainly I was in the early '70s, where there's a certain set-up, the musicians are out there, the engineer there, the producer's there running the session. And then suddenly all these punk bands turn up, they're running the session from out there, you're there as engineer, and the producer's chair is empty. How much guidance dare you pipe up with when you know things are not right? And Killing Joke had no experience [of producing] they all wanted everything loud and the four musicians would come into the control room for playback, which they would want at a pretty deafening level and they wouldn't sit down and say reasonable things like "could we turn the bass up here Phil", they'd wait 'til the playback was going on at full level and come and stand at the opposite side of the desk and scream at me. Scream into my face, each and every one of them about "my keyboard's got to be more distorted" all this kind of thing coming at me, and you're frantically [doing what you've been asked] and then along comes the next person who wants something more extreme on their instrument. Man, oh man, it was very difficult, everyone was just piling in. That's why for me even if it's from within the band there's got to be someone who's in control, some sort of team leader. I don't think many people work like that now but a lot of that did go on.

I worked with The Clash and the first thing I did with them was to mix '(White Man) in Hammersmith Palais' [in 1978]. It was quite clear to everyone in the studio that Joe and Mick were the leaders of the band. If the drummer did turn up he was perfectly happy, when the bass player…[sighs]… we'd work really hard the day before with Joe and Mick arguing [laughs] but getting to the point where they're both happy. In walks the bass player the next morning "Where's the bass!" and really complaining, I'm pretty sure he walked out of the studio and slammed the door. Did we turn the bass up after that? We certainly reviewed it, it wasn't as though me, Joe, and Mick ignored him. But it wasn't a collaborative discussion, it was a screaming match.

RW: Do you do things differently when the artist is out of the room?

PH: You do if you know it's a mistake. Classically Pete Waterman would come in and ask where the kick channel was, it was nearly always channel One and he'd just go over to it and push it up another dB, but whether you leave it up that dB for the final run off [knowing pause] I don't know?

Lamont Dozier and Cliff Richard

Having worked at the Marquee Studios when Pete Waterman had bought one of his Motown idols Lamont Dozier (of Motown songwriting legends Holland Dozier Holland) across from America, Dozier had got to know Phil Harding and several decades later Phil was asked to mix an album that Dozier was working on with Cliff Richard, which was released in 2011 with the title *Soulicious*. Waterman described Dozier as "My all-time hero" (Waterman 2000, p.127). He admired how he worked recalling that "He said 'I only ever wrote one song. I just did it 66 times'. What a wonderful man" (2000, p.128) (we would agree with Dozier about his 'one song', but perhaps not quite in the way that he meant it).

PH: Before Waterman got together with Stock and Aitken he had Lamont Dozier come over for six months, so I worked with Lamont and Pete at the Marquee on projects that largely didn't surface, the only thing that was successful from his visit was the song he wrote for Alison Moyet. I wasn't involved in that although Ian Curnow had played on the demo. It was a fantastic period. What was fascinating, unusual and almost frightening for the acts was that during a vocal session, he would be wanting a vocal change and here was a person capable of hitting the talk-back button and singing it fantastically down the talk-back mike and everyone in the room was like "Wow, if only we could sing it like that!" [laughs]. So, I got friendly with Lamont and stayed in touch. And I got a call in 2011 to come out [to LA] to mix the Cliff Richard album [Soulicious]. He'd produced 11 tracks and his feeling was, "this is for the UK market so what engineers do I know who would come out and mix with me" and I got a strange email from David Gest. He instigated the album, the story behind it is that he went to Cliff and his management and said "look, I've been working with all these soul artists, the Temptations, Candi Staton [etc.] and arranging tours in the UK for them. Why don't we do a collaboration where it's duets between Cliff and these artists?" It was quite strange for me, I'd worked 'with' Cliff Richard a number of times but never spoken to him, never met him, dealt with his management company. Eleven tracks were being mixed and we were in Los Angeles and he in his New York apartment at the time, and every night we would FedEx a CD to New York, he didn't want digital files [which would have been easier to send], and getting emails back the next day, which actually were Cliff's words but from his assistant, and the only comments were about vocals, whether it be him, on the backing vocals or the other vocalist.

Connection to Motown and Other Hit Factories

RW: Did you feel that connection at PWL with Motown?

PH: Almost literally the day we arrived there, and there was myself, the maintenance engineer, and either one or two assistants, but all had left the Marquee to go and work there and basically had a team meeting saying "the only person I'm paying a wage to is

Sheri who's answering the phones and working the reception, and the rest of you guys coming from the Marquee are all becoming self-employed from today onwards. I'm not paying you a wage, but I can promise you we are going to work hard to bring the work in, and we're calling it PWL Empire and we're going to try and become the Motown of the UK". They may not be those exact words but it was something like that, a motivational team leader speech at the beginning of the whole thing.

RW: Does the X Factor machine feel a bit like the PWL System?

PH: In the early days Simon Cowell might have thought he was modelling it [on PWL], he made a big claim of Waterman mentoring him. But I'm not so sure, because the difference there is that there was never just one production team behind an X Factor artist. Simon learnt that you can't be reliant on one production team. He was always less hands on than a Waterman or a Tom Watkins and very good at delegating, it would be rare and something drastic that would make him go to the studio. [...] He's done a good job of his label by delegating well since the late '90s with teams of A&R people and different production teams. Looking at the X Factor artists' albums there will be three, four, six production teams, which for all the production teams and songwriting teams around is great. Many people get a stab at those projects [and] it's a good way to utilise the singer's time because whilst one production team is working on four tracks the vocalist is working with another team.

The X Factor model has similarities with regard to its 'hit factory' nature, which Motown and PWL share, but it differs by not being 'the band' in the way that the team at PWL were, or the legendary session musicians The Funk Brothers at Motown and the songwriting teams of Holland Dozier Holland or Stock Aitken Waterman. The X Factor model of Simon Cowell uses different 'bands' (production teams) for efficiency of artist's time but also so that, if one production doesn't work it can be sent out to another team to see if that works better. Its factory model is more distributed, less like the PWL band, and maybe one reason why it can be sustained longer given it will not have the same intensity of group collaborations that an integrated or family model may have.

Solo Album Material

CJ: When you were working on your solo album, The Story of Beginners (2012). How was this different to when you are working on your own material when you don't have the team leader?

PH: You're talking about my solo album that I put out in 2012, oh gosh!

CJ: Having worked in such an intense collaborative environment how did it feel being just solo?

PH: It was a nightmare [laughs] [...] in terms of my writing prior to that [solo album] it had always been keyboard based and driven by great musicians like Ian and I would feed in ideas and help guide things'. [...] And we went into, what I considered jokingly, as a sort of Elton John and Bernie Taupin kind of collaboration, I would encourage lyricist Mila Bogen to write out in the format that I'd taught her, four lines for a verse, four lines for a chorus, etc. and I would put music to it and that was inspiring for me, to meet someone that could deliver words that I was enjoying and I could sit down with a guitar rather than with a keyboard and all the technology, and that bit was great, but then, as you are asking, producing myself, there were enjoyable moments. It was good recording other musicians but when it came to recording myself on vocals it was an absolute nightmare. I didn't feel I had the courage enough, because of my vocals, to ask someone to come in and produce me. It was quite a problem and that's why I've done nothing [solo] since. I did it to fulfil a creative period rather than wanting to sing vocals. I wouldn't recommend producing oneself to anyone!

Phil had returned throughout the interview to his observation that a team leader is essential. Yet, understandably, as he says for reasons of lacking enough courage to ask someone else to produce because of lack of confidence in his own vocals, he goes against his own advice and hence finds producing himself 'a nightmare'. This is tinged with humour throughout, it's not the same nightmare that the chaotic Killing Joke album experience was, but it highlights the same issue from another angle. He has taken his own advice more recently though with a project with the 24CLUB, a former band he was involved with under different names.

PH: Recently I've been involved in a collaborative project, an EP by a band called the 24CLUB, which was not their original name, late '79 early '80s post punk new wave period. We've come back together recently to revisit some old songs and do some new ones. I'm on bass and backing vocals.

I have nominated the singer/guitarist in LA to be team leader, so that they have the final decision. We've had some differences in chords for example and I have gone with his decision. It's been a tricky coproduction, somehow jointly doing technical decisions remotely has been difficult. We had to get it done quite quickly because some tracks were for a film. But it was difficult, and there certainly won't be another Phil Harding solo album [laughs].

Perhaps the solo album experience is part of the reason why we were keen to interview Phil Harding. By his own admission he is not a 'solo' operator, he is a team player, a coproducer. He is enormously good humoured and generous with his time and knowledge, the kind of personal traits that may not put them in the spotlight, but without whom there would be no spotlight, nor anyone under it to shine.

7 Group Genius, Scenius, the Invisible, and the Oblique

Eno, Lanois, and Communities of Creativity

Robert Wilsmore

Introduction

In his talk for pressure group Basic Income UK in King Square, London, in 2015, Brian Eno articulated his thoughts on the relationship of genius to the 'scene' in which it is located, saying:

> Although great ideas are articulated by individuals, they are nearly always generated by communities. [...] All these people who are called genius, actually sat in the middle of something that I call Scenius.
>
> (Eno 2015)

The naming of the individuals that rise to the top to be credited with authorship is the creation of the 'genius' artists, and upon their naming the others, the unnamed, fall away so that all that is left is the one name (Picasso, Stravinsky, etc.) to whom the credit is given. For Eno, the notion of Scenius is aimed at bringing the unnamed back into the frame, to promote and to credit the community that coproduced the art. His notion sits comfortably with our leaning towards the rhizomatic philosophy of Deleuze and Guattari and their statement at the opening of *A Thousand Plateaus* that "Since each of us was several, there was already quite a crowd" (1987, p.3), and for us Scenius is the combination of people, time, and place in which the pop music multiplicity of Toast theory exists.

At the start of Keith Sawyer's book Group Genius (2007) he notes the collaborative efforts of the Wright brothers in developing the first plane to fly, but then immediately follows this up with a communal development that does not have the author–genius names attached, that is, the development of the mountain bike. Although aspects of contributions can be identified with regards to the mountain bike's evolutionary creation, Sawyer's example also notes the state of production teams where "many creative collaborations are almost invisible – and it's these largely unseen and undocumented collaborations that hold the secret of group genius" (2007, p.5). The operations that result from these multiple and unidentifiable contributions he calls 'invisible collaboration'. Similarly Eno, speaking at the Sydney Luminous Festival in 2009, said:

> Scenius is the intelligence of a whole operation or group of people. [...] Let's forget the idea of "genius" for a little while, let's think about the whole ecology of ideas that give rise to good new thoughts and good new work.
>
> (Eno 2009)

Both Sawyer and Eno, Bennis and Biederman (in *Organizing Genius*), and Deleuze and Guattari, are making visible the necessary contribution of the many that are lost with regard

DOI: 10.4324/9781351111959-8

to their part in the production. An objection is that they themselves are named (Eno, Sawyer, etc.) just as the traditional genius is named, but as Deleuze and Guattari answered "Why have we kept our own names? Out of habit, purely out of habit. To make ourselves unrecognisable in turn" (1987, p.3). The concepts and philosophies of Scenius, Invisible Collaboration, Organizing Genius, Rhizomes, and Toast are not to eradicate the name of the genius, but to establish that name as a collective noun that holds the productive community within it.

Working with Eno

Eno's coproducer on the U2 Grammy award winning album *Achtung Baby* (1991), Daniel Lanois, made explicit a simple approach to their collaborative working model, writing "When working together, we have a simple agreement – I agree with everything he says, and he agrees with everything I say" (Lanois in Savona 2005, p.128). Similarly, in an interview with Massey he notes that "without speaking a lot of words, we just walk into a room [and] in a matter of minutes there's something happening. [...] That's much more important than suffocating the subject matter with too many opinions" (Massey 2009, p.23). Lanois worked in familial and integrative fashions as a producer with U2. Although he was not 'in the band', a key strategy of working with others was to integrate into their contributive element, in this case by being 'one of them' as a musician and performer. He writes "Whenever I work with U2, whether in Slane Castle or the rented house just outside Dublin where *Achtung Baby* was recorded, I try to become a temporary member of the band. I usually like to plug in a guitar or play a piece of percussion. It gets me involved in the arrangement and increases the adrenaline" (2005, p.127). Likewise, in an interview with Massey, Lanois said "I like to be part of the band, especially during the first stage of a record where you've got to juice up the room" (Massey 2009, p.22). There is a familial approach to his strategy whereby roles become fluid and dynamic, although his overriding position is one of being a producer and not band member, the description that he is a 'temporary member of the band' can be viewed as being a 'temporary member of the family'. Ultimately, this is a tactic used to influence the strategic outcome, that is, to influence the arrangement and inject 'adrenaline' into the system, presumably so that this energy comes out in the recording. Lanois was the principal producer for *Achtung Baby* with Flood (Mark Ellis) as engineer and Brian Eno operating as a routine disrupter and challenger, letting the process continue without him and then coming in for a week after not being involved for a couple of months (Eno 1991). In a complementary fashion he offers his 'Merlin service' to the production team, as Burgess quotes Eno in his Merlin category "Normally I don't stay with the project for the whole time. I deliberately keep out so that I can come back with fresh ears" (Burgess 2002, p.10). In our interview with Bruce Woolley, discussing the Grace Jones album *Hurricane* (2006) he worked on produced by Ivor Guest with Eno as consultant producer, he recalled that "I remember Ivor Guest at one point just despairing because he said "Oh no, Brian's coming in tomorrow and all he's going to do is say the exact opposite of what I want to do." Intervention and disruption are part of Eno's strategies for finding the new, for ensuring things do not stay the same and hence prevent stagnation and increase creativity. Perhaps the most reported example of Eno's intervention in collaboration with Lanois was the myth (or at least the exaggeration) of the recording of *The Joshua Tree* (1987) opening song 'Where the Streets Have No Name' that Eno became so frustrated that he tried to erase the tapes, is only true in thought it seems. As Eno later clarified:

> My feeling was that it would be much better to start again. I'm sure we would get there quicker if we started again. It's more frightening to start again [...] my idea was to stage an accident to erase the tape so that we would have to start again. But I never did.
>
> (O'Hare 2007)

That starting again is 'frightening' is balanced by Eno's excitement of being scared; using Eno and Schmidt's *Oblique Strategies* as a way of exploring Eno's collaborations, Marshall and Loydell subhead a chapter section with the card "Do we need holes?" to demonstrate Eno's technique that may lead to the erasure of the work of collaborators (as well as his own) and what this does to the creative process. Erasure can lead to a hole, which leaves "a void to be willingly and enthusiastically explored. Or in aural terms, a musical hole to be filled" (Albiez and Pattie 2016, p.178), their discussion leading on from Eno's comment on collaboration that "when you work with somebody else, you expose yourself to [...] the risk of being sidetracked", which for Eno creates the conditions that he wants to be in place such that it "will take me into new territory". An emerging theme of Eno as a collaborator is someone who is prepared to break the production in order that the team then have to 'start again' and in doing so move to a more creative place than was previously arrived at. The community that is the artists and producers is provoked into being a creative community rather than a repetitive one. One of the services provided by this particular Merlin is to go in, punch a hole, and allow it to refill in a different way.

As with the observation from Lanois that he became a temporary member of the band, so Eno is described by Eric Tamm as a 'Ghost member of the band' with regards to Talking Heads with whom he coproduced *More Songs About Buildings and Food* (1978) and *Fear of Music* (1979), comparing him to the relationship of George Martin and the Beatles, the 'quintessential' (fifth element) as we put it at the start of this book. Tamm describes one of Eno's working methods that he had developed on his own progressive works as well as when collaborating with David Bowie, that in recording sessions he "kept a few 'spare' tracks for himself on the reels of twenty-four-track tape [...] On the spare tracks Eno would feed in this or that instrument and treat it electronically" (Tamm 1995, p.160). Hence Eno had dug himself a hole, not in the sense of boxing oneself into a tight corner, but that punching holes not only allows others to fill them, as Marshall and Loydell put forward, but also that he creates holes so that he can fill them, as is the case Tamm presents with the Bowie and Talking Heads examples. The digging a hole metaphor is reversed from being one of being hemmed in, to being one of taking control, of making productive space.

In Tamm's Book *Brian Eno: His Music and the Vertical Colour of Sound* (Tamm 1995) his chapter 'Collaborations' focuses on Eno's work with three particular artists with the subheadings 'with Robert Fripp', 'with David Bowie', 'with Talking Heads and David Byrne'. They detail different aspects; on Fripp the section focuses on describing the music that the collaboration produces, with Bowie the issue of entangled authorship is forefronted, and with Byrne the issues of ethics of cultural appropriation, the cultural philosophy of McLuhan's global village and the influences of Stravinsky, Debussy, Elvis, and the Beatles all come into play. We see an exploration of technical work, joint authorship, and philosophy that are all at play in the collaborations. On working with Bowie, Tamm writes "According to Eno, Bowie was interested in working with him because he found that his own creative ideas were running out" (1995, p.157). It is a familiar theme, one echoed and exemplified in Brian Wilson's response to Howard Massey that he needed coproducers in order to bring new ideas. On the question of joint authorship, Tamm writes on discussing Eno and Bowie's' work on 'Warszawa' from *Low* (1977) that:

> The whole matter of authorship is complicated in collaborations of this sort. Often, a range of duties seems to be shared by a number of musicians: when there is no neat division of roles (composer, arranger, instrumentalist, producer), but several people are active in all of these spheres at once, then whose piece it is may boil down to who is paying for the studio at the time.
>
> (Tamm 1995, p.157)

The passage neatly sums up the problem of a need to identify 'who' is the author. It seems that we do not like it when we can't do that, and in this example, ownership (which equates to authorship here) is given to the person with the highest ranking status, which in this case, as so often in life, is the person with the money (material capital leading to cultural and social capital). This returns us to why Eno put forward Scenius as a concept; it would be so much easier to 'give' authorship to Bowie, or Eno, or both, and if we could identify the complementarity of the situation, the 'composer, arranger, instrumentalist' etc., we could feel confident about knowing who did what. But this "complex authorial situation" (Tamm 1995 p.157) requires a different approach. Hence, the acceptance of the one and the many that can be described by Integrated, Familial or Distributed Collaboration, Scenius, Rhizomes, or Toast, ought to ameliorate the problems arising from accreditation to the one, or more specifically, the one with the money. Why is the latter bad? Because it maintains that hegemony, its continuation, allows a dominant model to keep control over a creative scene rather than allow the creative scene to own production. It is a model that might benefit from having some holes dug in its territory.

Lanois, in terms of the 'scene' in which he is located, sits somewhere at the end of modernism and the move through postmodernism. That is, that whilst he recognises the moments that are produced collaboratively, he also sees the continuation of the individual genius, saying that "you have to remember that there's another Stravinsky out there, another Beethoven, another Mozart" (Massey 2009, p.21). When Massey asks what Lanois meant by complex music being a network, he replied "if you and I are playing instruments, I start playing a certain way, and then you'll accommodate my way of playing in the way that you play, out of respect for being harmonious and trying to get good results", here he notes a classic 'give and take, take and give' approach to improvisation amongst the group. He describes working with Peter Gabriel and David Rhodes as a joint construction process where they even wore construction-worker hard hats in sessions to humorously emphasise the work-in-progress nature of the composing and recording scenario. Lanois contrasts this with working with Bob Dylan on *Time out of Mind* (1997) where the group authorship in this context comes from the sound "We put 11 people in the room with him [Dylan] [...] There's an automatic depth of field you get by having 11 people playing together" and he concludes fairly straightforwardly "If the songs are written beforehand, you do it one way, and if it's more jam based, you do it another" (2009, p.22). But clearly, like Eno, Lanois is excited by the unknown. He likes to keep recording in-between takes to capture the creative moments that just happen in the moment:

> There's a momentum that I tap into that's usually based on the thrill of an idea. It doesn't matter where the idea comes from; what's important is that you chase after it. And if a result comes from an idea, then that usually triggers a second idea, and a third one, and you start capitalizing on the potential of the momentum in the room. There's a power that exists when you have a group of folks in a room. You sort of go down this labyrinth that you might not have expected beforehand.
>
> (Massey 2009, p.22)

This labyrinth is not one that is a complex set of pre-existing routes but rather that each new path chosen then leads to further unidentified choices that eventually lead to the goal. In Lanois' case, this is the 'juice' that leads to those 'Whoa!' moments. To connect Sawyer's 'sceniusic' mountain bike development to the sounds from the studio, he ends his opening chapter (that began with the aeroplane and the mountain bike) with the W.L. Gore & Associates company whose virtually non-hierarchical structure kept 10 per cent of employee time for them to speculate on new ideas. One small cluster, realising the cables on their bike brakes coated in plastic film might translate elsewhere:

> Realised that by putting a similar coating on guitar strings they could prevent the dulling of sound that occurs when natural oils from fingers corrode the strings. [...] Soon after its release Elixir quickly became the top-selling acoustic guitar string – a success that emerged from improvised innovation.
>
> (Sawyer 2007, p.19)

We never have to look very far from what is in front of us in order to see the impact of the connections beyond it, but the point is that we must look. Eno looked and took a slanted, oblique line of flight that dug holes and changed the object, but it is only right to follow that up with 'Eno' as the collective noun for the scene, the zeitgeist, the postmodern milieu. Call it want you may, Eno is a crowd and that crowd changed music.

8 Grace Jones, Spontaneity, and Collaboration in the Moment

An Interview with Bruce Woolley

Christopher Johnson and Robert Wilsmore

Bruce Woolley is a songwriter and producer best known for his collaborations with Trevor Horn and Geoff Downes on the song 'Video Killed The Radio Star' and with Grace Jones on 'Slave to the Rhythm'. Outside of the mainstream hits he is widely respected for his work with The Radio Science Orchestra, which he co-founded in 1994 with Chris Elliot and Andy Visser, and in particular for his love of the theremin. At the time of our interview Bruce was putting together preparations for the 100th anniversary of Leon Theremin's creation, albeit, it seems, experts debate exactly which year the theremin can be said to have been invented. Towards the end of the interview we were joined by another collaborator on the theremin project (and inventor of the 'Theremin Bollard'), David Young. We have kept the interview as a straightforward transcription to capture the spontaneity and ephemerality of the moment.

Bruce was the invited keynote speaker for the Music Industry Day at York St John University on 22 January 2019. His talk took us through his life in songwriting and production, and in the afternoon, he held sessions with Music Production students in our studios, his generosity and humbleness shining through at all points. In the keynote speech Bruce included in his top five tips to all new producers that they should "Collaborate. Find the people who can do what you can't".

In keeping with Bruce's concern to make sure the students got as much out of the day as possible, he went round the room in which we held the interview, which was attended by Music Production postgraduate and research students, with genuine interest asking them what their main focus was on; for most it was a mix of engineering, producing, songwriting, academic writing, but with a clear preponderance on songwriting, which Bruce picked up on. The interview segued seamlessly from Bruce's natural inquisitiveness of the postgraduate producers, picking up in particular on Chris Readman, noting that the roles of songwriter and producer are not boundaried anymore.

BW: When people say songwriting I wonder what that means now? Because, [and] we talked about Bacharach and David [Bowie] earlier and obviously the Beatles and going back as far as Noel Coward (and people you've probably never heard of), and that's what you'd think of as songwriters. But of course, it's very different now. One of the big things we talked about was finding the people to do the bits that you're not as strong at. I used to do songwriting in the days when people would thrust you with a completely unknown song[writer], and the publishers loved doing this, and you'd meet this guy for the first time in a room and you're supposed to write a song, and you'd find out that he actually couldn't write lyrics, and I can't write lyrics, so it was like "so what the hell are we going to do then? How are we going to do this?" So that 'yin and yang' is crucial.

RW: In the early days you said that you were paid £50 a week to write songs.

DOI: 10.4324/9781351111959-9

BW: Yes, it was a huge sum and it allowed me to live in London. [to get this type of work] I had to move to London, there was no option, I needed a job and I got this job as a songwriter, and they put an upright piano in this office and I used to go in every day and write songs, and that's how it began really. It was a very Tin Pan Alley, 1950s approach.

RW: Did you have to write a certain number of songs a week?

BW: It didn't work like that, the publisher needed to be entertained actually, but he definitely needed songs, and the contracts were terrible, really. To give you an example you have a 50/50 songwriting deal with the publisher, which these days is not good, but then I realised that the 50 per cent was being collected in America and then sent to London where they would deduct a further 50 per cent, so the songwriter was only getting 25 pe cent. So, read the small print [...] get a good lawyer, it's worth it in the end.

RW: You cross all the boundaries – performer, writer, and producer, and you co-write and coproduce. Starting with songwriting, you've written with Trevor Horn and Geoff Downes, with Grace Jones and others, can you talk us through how you co-write? Just a minute ago you were talking about a collaboration where neither of you could do the lyrics and it wasn't going anywhere.

BW: Yes, that's depressing. You want to avoid that. So, the first thing is to be working with someone who does what you don't do. Grace [Jones] is a great example because I met her on the Slave to the Rhythm sessions where, from her point of view, I was just the programmer or co-writer, and she didn't really know who I was. She and I just found ourselves in the studio together and there was a drum machine and a keyboard, and we just started knocking stuff out. It was New York and it was the '80s, we got on very well, and although she can be frightening to look at and can be quite frightening in person, she's a sweety, and I like to think I can recognise good lyrics and can recognise a person who has the gift of writing lyrics. It's all very well saying write lyrics, I can write lyrics, but I know perhaps a handful of people that genuinely have that gift. And it's like, we're in the room, a blank sheet of paper, and they're throwing words at me that I would never dream of myself, and they probably hadn't thought of them until they were working with me or somebody else. So, it's a spontaneous, almost supernatural, thing that happens when you get a lyricist and a musician together. And you can't make that up really. That began to happen with Grace quite spontaneously, I wasn't expecting it at all, and we went on to write a whole album together after Slave to the Rhythm.

RW: So, you had a drum machine and things began to happen?

BW: It was a TR808, it was the early days of Midi and I had managed to hook up a Juno 106 to the TR808 and that was the basic rig, and she said [attempts Grace's accent] "Bruce, I have this idea for a song, it's called 'Party Girl', and she's completely out of it", and I thought "that's good, I like that" so I hit the TR808, well it's a bit like a typewriter in a way and you set it into write [mode] and then you could just tap the beats in [beat boxes a demonstration] and Grace immediately liked it, so I played a couple of chords literally [sings to demonstrate] and wrote them into the MSQ 700 which was a very early digital sequencing recorder, and she was like "Oh my god" and that's how it started, just like that, we had suddenly just started writing a song.

RW: So, where does it go from there?

BW: It's a bit like having a baby or something, a bit like watching a cell reproduce from nothing into something, and anyone that's felt that with another songwriter would understand that you're really creating a new life. That sounds a bit pompous doesn't it, a bit pretentious, but I think that's kind of like what it is. And there are many other ways of writing a song, someone can give you a lyric and you can go away and think about it, or you can give them some music and they can go away and think about it, and they're all valid, but there's nothing quite like that spark of being in the room with somebody and just creating something from scratch. That was 'Party Girl', and that was the first song that she and I wrote.

[But] at that point the Slave to the Rhythm sessions were getting quite dark, there was a lot of doubt around, and I was thinking that we could put 'Party Girl' out as the single instead but politically that was impossible at that point. And so Grace and I kept on writing after the Slave to the Rhythm project and we worked on her next album with Nile Rogers for EMI records.

RW: May I stay on 'Party Girl' for the moment? It starts in the space together evolving on its own between the two of you, then did it continue like that, were there times where you went away and did separate bits and came back?

BW: No, whenever I worked with Grace there was virtually none of that "let's go away and think about it", we did it there and then, and that's where the staying up all night thing came in because she's nocturnal [laughter], and I couldn't get out of it. I couldn't say "Grace, look I'll see you tomorrow", I was there, and I had to keep going, you had to keep going until the song was finished, and that's how she operated. She would come in with titles sometimes and we would start from scratch from a title, in fact that's often how it began. It definitely was not protracted because she's the sort of spontaneous person that you have to go with it, some artists are like that, in any production situation, any studio situation, you cannot put it off to another time, you've got to do it there and then otherwise you won't catch it, it'll disappear, it's ephemeral.

RW: There weren't revisions afterwards?

BW: Not really, no. But lots of songs have been written like that. It's not unique.

CJ: Does that include all the arrangements and sounds as well?

BW: Once we had written the song I used to walk around with a [Sony] Walkman cassette and we'd record it on that, and then we'd regroup and record it with her doing a vocal and we'd put in other sounds and do an arrangement, but based on that 'day one' writing session.

CJ: It's interesting that the initial stimulus [of 'Party Girl'] was the 808 and the beat that came from that rather than a chord sequence or a lyric.

BW: As you say, the beat was there first, and if the beat was not there first then Grace wasn't interested. The beat had to be there, there was no way you could intellectualise it and say, "trust me it'll be great", if she's not feeling it straight away then it's not happening. So that's why the 808 drum machine approach worked really well with her.

RW: Going back to The Camera Club and working with Trevor Horn, and the songs 'Clean, Clean' and 'VKTRS', the two songs that appeared on both The English Garden and The Age of Plastic.

BW: When I first met Trevor he was the house producer at this 'Tin Pan Alley' publisher and he and I had never written, he was just the producer, [...] the poor publisher ended up getting these studio bills through that were off the scale [...] and he went out of business, so we were all out of a job for some months. So, Trevor and I hooked up after the dust had settled and realised that we could write songs together, not having done so before, and 'Clean, Clean' was that first song. That was the first time I realised that he was a lyricist, (as I was talking about earlier), and prior to that I had been writing quite middle of the road material, like Hall and Oates, trying to get a song for Cliff Richard or Rod Stewart, trying mainstream stuff. But moving to London in '76 there was a lot of energy, the whole punk thing was kicking off, and 'old American music' looked like it was almost over at one point. It looked like the end of what was regarded as song-writing and music 'cause the punk thing was such a powerful explosion in London, it's hard to explain now but it really did shake things. It didn't last that long, but it made a cultural impact and Trevor and I picked up on that and our first song 'Clean, Clean' was a strange story about future warfare and it was like actually no one's probably written a song like this before. It was probably influenced by Kraftwerk or The Stranglers (you're always influenced by what's going on around you, in whatever era you are), and so that was a discovery.

RW: Your writing methods, was it together in one room?

BW: Yes, just an upright piano and a guitar, so everything was written like that. Trevor, rather annoyingly, never wrote anything down, so not only did I have to do all the music, the piano and the guitar, I also had to write out all the lyrics as well. But I have got a lyric book full of that early stuff that makes for amusing reading.

RW: Mike Hirst was your producer for English Garden (1979), and both 'Clean, Clean' and 'VKTRS' were also on The Age of Plastic which Trevor Horn and Geoff Downes produced. Was it an issue for you that you were not a producer on your songs at that time?

BW: That's really complicated, it's complex because we were in a DIY, do-it-yourself punk era, but none of us had had any success so we always thought that everybody knew better than we did. It's one of those schoolboy things "well, he's done this so he must know what he's talking about". Mike Hirst had had something like 35 hit singles and was a very successful record producer at the time so we thought he must know what he's doing, unfortunately I don't think he did and was just reaching the end of his career, bless him! It was a real shifting of eras, it was just before the '80s, and he was quite rooted in the 1950s actually. He was into rock n roll and didn't really like punk music, he tolerated us because he thought he was going to make some money and was a nice guy, but probably not the right producer for that album. Perhaps if we had done it ourselves as engineers, we would have got a better result, we might have been better off with someone like Mike Chapman who did the Blondie stuff, and prior to that as Chin and Chapman, a huge English pop production team. And Mike Chapman went over to the States [and] really got it right for Blondie. If you listen back to some of their early demos and arrangements, like Heart of Glass, it's almost a different song, he was responsible for their success. So unfortunately, I think we were with the wrong producer, which might account for the fact that we never had a hit, seriously, and the labels in those days, if you didn't make the Radio 1 playlist, then that's it you're over, finished. What's good about now is you can always be rediscovered or can always make new material and it's going to get found if people want to find it. You can make music now and it can be discovered, that's a wonderful thing about now. That's a healthy development.

RW: Thomas Dolby was in the group, did you write with him as well?

BE: Not as much as I would have liked to. Basically, Trevor, Geoff, and I were in a group together and then we split up, we'd already written 'Clean, Clean', they went their way and I went my way with The Camera Club and Thomas Dolby, and then we wrote 'Video Killed the Radio Star' in this sort of split-up state as it were, so we both wanted to do the same song and I wanted to do my version, which was a bit more straight (not as good, by the way, as The Buggles version) and they did their own version, which I'd helped them develop, incidentally, and Thomas Dolby ended up playing a lot of the keyboard lines that Geoff Downes had written in the arrangements of the songs when we'd written them. So it was, kind of, a bit messy actually [laughs]. Thomas was great though, a great, great, player! And obviously went on to have great solo success. So, there was a conflict there and an awkwardness with the record labels because I was with CBS and The Buggles were with Island [Records] and The Buggles had the big hit and I didn't really. But I went on to tour America and got a bit more, in those days we called it 'credibility', I had loads of credibility but I didn't have the hit record [laughs].

RW: You came back to work with Grace Jones, about 2006 you started work on the Album Hurricane.

BW: Actually, I started work on Hurricane in 1991. I spent 20 years working on it, but you weren't to know that!

RW: So how did you start writing?

BW: We just started writing again and came up with a song called 'Love You to Life' and she was working with Chris Blackwell again and she went through all sorts of different incarnations with her labels and at one point was signed to an internet label, this is way before iTunes and it was a novel idea that you could release music on the internet. So, the whole thing took about 20 years. Then Hurricane came out eventually.

RW: Did 'Love You to Life' change over its 20 years gestation?

BW: It did go through a lot of incarnations.

RW: Some of your children come along very quickly and some take a long time.

BW: That's so extreme, isn't it, when you say it like that. But the germ of the idea probably came together quite quickly. But I didn't get so involved in that album as she had other collaborators, Ivor Guest produced it and Brian Eno was working on it as well, he's an interesting guy.

RW: He is noted as being consultant producer as opposed to 'producer', was he someone that just added ideas?

BW: I don't know if anyone can figure out Brian Eno at all? [...] He's a great one for doing the opposite, I remember Ivor Guest at one point just despairing because he said "Oh no, Brian's coming in tomorrow and all he's going to do is say the exact opposite of what I want to do" [laughter]. He's a lovely guy by the way, a real gent.

RW: We are talking about songwriting and producing interchangeably. The line between writing and producing is blurred now, is that how you see it?

BW: It really is, that's very true now. It's very different from the Tin Pan Alley days where you write the song so that the artist and producer does it. That's way, way, old fashioned, although there are still occasions when a writer can submit a song to an artist, you can still do that. There aren't that many artists that don't write their own material but the two are so entwined now, the production and the writing of the song because they tend to happen at the same time. It's much easier now for people to make things sound good because we've got all these great tools. You've got Abbey Road on your laptop really. Literally, if you look at the echo chambers and some of the plug-ins they've got going. So, it's a very different time. I still think a lot of artists would benefit from an external producer, but that job is not what it used to be. I'm sure even Trevor would agree that, I think producers need to 'self-commission' a bit more. I know a lot of people from my generation who probably aren't as busy as they should be, because they are still thinking in those terms. The thing is with technology and the analogy with 'Video Killed the Radio Star', is that of the blacksmith seeing the Model T Ford coming into the village and thinking "That's me out of a job" but the trick with the blacksmiths is that they started to learn how to mend motorcars and stock petrol and stay in the travel business. The blacksmiths turned into garages, and I think we have all got to learn how to adapt at any moment actually to some new thing that comes along. The role of producer has changed dramatically.

[We are joined at this point by David Young, producer, composer, and creator of the Theremin Bollard.]

RW: [to Bruce] Something about technology clearly fascinates you, for example your use of the theremin. In the '90s you started working with The Radio Science Orchestra and writing for film. Was that very different to writing pop music?

BW: Every now and again you get these epiphanies and the theremin was a real shake up for me. On its own it is a limited voice but there's something about it, almost a spiritual aspect. That doesn't sound too pretentious does it? And that put a spin on my activities, and we used it in tracks in a couple of films, notably The Avengers (which probably gets a lot of 'rotten tomatoes' in reviews!). We got Grace Jones to sing the title track and we recorded the strings at Abbey Road and funnily enough a lot of the players on that were on 'Slave to the Rhythm'. Working with film is very different to working on a track that is to be played on the radio or in a club. The Radio Science was a bit of a departure for me, a bit of a life saver for me as I was getting very disillusioned with pop music at that point in the early '90s. Pop music wasn't really doing it for me. I went back into exotica, Martin Denny, Les Baxter, Raymond Scott, one of Bob Moog's inspirations, and the theremin still speaks [now] and is on the increase in popularity, and we have the great inventor with us, David Young. [to David] Does everyone here know all about the theremin?

DY: Oh yes! [much laughter from audience, all of whom have been taught by David].

RW: The theremin has become known as being the past's version of the future. Did that make it a 'one trick pony'?

BW: It does have a stereotype, people always say it was from black and white B movies, but it wasn't in that many.

RW: In the studio is it an instrument to record like any other instrument?

DY: It's a pretty straightforward thing to record on the whole. If it's one it's OK, it's when there are multiples that it gets problematic. It's very singular, it's a monophonic instrument, it has a particular tone, unless you start processing as I do, but it sounds like the voice and Lydia Kavina, the great niece [and pupil] of Leon Theremin taught me and always said "sing it in your head. It's the voice in your head". Some of you may have heard Charlie Draper playing "The Great Gig in the Sky" from Pink Floyd's Dark Side of the Moon on a theremin. An absolutely stunning piece of Theremin playing.

BW: [And] Pink Floyd were a good example of taking a beat group format of an organ, drums and bass and guitar, but then made music that was very different to their contemporaries, it wasn't pop music as such, so I think there is always an angle. It [pop music] will change, it's just that we've reached a plateau right now and we're not sure how we're going to break out of that. [to audience] Is it a valid argument that there is a certain homogeneity attached to the fact that we all have the same equipment, we have the same plug-ins, same Apple loops, we're all buying the same gear, so that's going to make things sound a bit similar and we need to break out of that somehow? And also, the way music is disseminated, now we can get anything at any time. We have explored the territory of sound so thoroughly [...] this is possibly the beginning of some fantastic new era but we can't quite see how that's going to happen, maybe a new technology, something like that [will drive the change]. You can always put a new angle on something.

9 Small Things of Value

Marginalia, Mental Health, and Coproduction (Abbey Road Case Study No.2, Part 1)

Robert Wilsmore

Like most terms 'coproduction' has different meanings and uses in different fields. In this book the term moves from what might be labelled as collaboration in record production in the first part, to the more general collaborative production of music that includes composing and songwriting. Amongst practitioners involved in working with mental health service users the term coproduction is often used in collaborative situations where an implicit ideology of equality is established (or intended) as part of the environment, and often where leaders also have lived experience of mental health issues. It addresses the problem in creative projects where a group may feel that they are 'over directed' and hence are 'done to' rather than being the doers. We have reflected on one particular project, the recording of the song 'Nothing of Value', in both this chapter and the next. This chapter has been written by someone inside the project writing as a composer, producer, and academic, whilst the subsequent connected chapter is written by an academic external to the project, Ruth Lambley, who is a researcher on the subject of coproduction in mental health services at Converge (described later) and who brings coproduction theory in that field to bear upon the project.

This chapter explores the production and the marginalia of the habitus of a song written for, and with, the mental health group Converge in the writing, rehearsing, recording and producing of the song 'Nothing of Value', composed by singer-songwriter and Converge music leader Esther Clare-Griffiths, with Chris Johnson and Rob Wilsmore, as well as members of Converge who contributed to the lyrics through working with Esther in Converge songwriting sessions. The reflection here focuses mainly on the creative and rehearsal processes and the recording session at Abbey Road Studio 2 on Wednesday 29 January 2020. Chris Johnson has outlined the Production Habitus in previous chapters including our first Abbey Road case study on the *Smoke Rainbows* project (with another mental health group Music Minds Matter). Here the intersection of fields and doxa are outlined between the power relationships and capital of the participating groups and individuals within them (such as York St John University, Converge, Communitas, and Abbey Road) and also to draw upon the marginalia of the environment, for example the effect of an apple in the studio on the proceedings compared to the effect of the name Apple on the participants and the recording. Although the minutiae of such marginalia are generally ignored because they are perceived as having little impact on events, their inclusion was of interest to us when trying to take in everything within the habitus of the field we were operating in. But rather than letting such things be insignificant, we wanted to note their value, which chimes with our project about the importance of feeling valued and valuing oneself however insignificant one might feel.

One key group of performers for the recording, Communitas, are a choir formed within the Converge project which has been running since 2009, and formed under the Converge Directorship of Dr Nick Rowe MBE and choral leader Chris Bartram. Converge provides courses at the university for adults who use mental health services, describing their main purpose as 'Education for recovery'. It has held to its five principles since its inception:

DOI: 10.4324/9781351111959-10

- To work together as artists and students.
- To build a community where we learn from each other.
- To engage and enhance the university and wider community.
- To provide a supportive and inclusive environment.
- Respect others, value yourself.

These five principles are an explicit doxa that, when put into practice, become the (structuring) structure that leads to the learning experience of, and benefit for, the participants. The 'Nothing of Value' project, without need to highlight the principles, went through these active structures. It is something of a continuing dilemma in this particular field, that of education for those with lived mental health experience, that such fixity has an inbuilt 'done to' quality to it. But it is one that is agreed to and welcomed in this case, as the once in a lifetime chance to record in Abbey Road brought the positive impact of cultural status to bear on the participants. Once achieved, that cannot be taken away, one's social capital may increase, certainly one's sense of achievement and value, of being a part of the story, not just an observer, is personally significant.

In collaborative terms the song was mostly composed in 'series', that is, a section is handed on to the next contributor who then adds to it. Generally, collaborative processes fall into three types: serial, parallel, and joint (see Alix, Dobson and Wilsmore 2010) where 'serial' is generally the process of one or more persons handing on their contribution to another, 'parallel' working is when contributions are worked on separately but at the same time (or rather, within the same time frame), and 'joint' refers to moments of direct communication working on the collaboration 'together', normally being in the same space at the same time (and increasingly together in an online space). Most collaborations will involve all three types, or variations of, these general categories.

To describe in simple terms the process for 'Nothing of Value', a short demo of the chorus line 'Are we nothing of value to you?' by Rob was sent to Chris and Esther separately. They work in parallel at this point. Esther returns a verse to complement it, although the tempo is much faster than the original chorus. Chris returns two almost fully complete demos of two songs. At this point the collaboration starts in series, that is, Rob passes on the chorus, then Chris and Esther work separately in 'parallel' responding to the stimulus (the chorus) but independently of each other. The responses are returned (handed on in series) to Rob who then makes unilateral decisions on what to keep and how to put them together. The songwriting collaboration has a natural hierarchy, as Rob proposed the project and put forward the chorus, he constructs the basic song from the three contributors. Esther's section (with some modification by Rob to gel it with the chorus) works as a bridge or pre-chorus, the chorus was sped up to meet Esther's tempo, and verse and other materials (such as 'middle 8' type material) were drawn from Chris' contributions. Esther's words for the bridge changed very little from the start. Chris, on the other hand, knowing the verse words were to be written by members of Converge, used his own improvisatory process of making up words to the tune as he sang on the demo, which is intriguing as the words sort of make sense, and yet they don't, something of a stream of consciousness approach to on-the-spot lyric writing. When the draft instrumental material was constructed, Rob did the same thing and recorded a version with similarly improvised verse lyrics, mostly for fun having been inspired by Chris' approach but also to avoid fixing the verse words so that these could be written by others. Esther then worked with her songwriting group from Converge to come up with the lyrics to replace the 'nonsense' ones. One final amendment to the lyrics came through rehearsals when the choir wanted something more proactive than the passive 'Are we nothing of value to you?' chorus line, and in the song changed the last repeating verse to the more positive and proactive statement 'We *are* something of value to you'. It was the right thing to do given the last sentence of the five principles of Converge is 'value yourself'. There was one short studio session with Chris, Esther, and Rob at York St John University where the structure was amended slightly through joint decisions made together, and Chris built this into the demo that would then be handed to the performers to learn; a necessary part of the preparation given time in Studio 2 was limited. It is clear already that some of the contributors, the lyrics writers, are unnamed at this stage and belong to the group identity of Converge rather than as individuals in the way that the others are named (Esther, Rob, Chris). Even within a project aimed at breaking hierarchies within the coproduction model, they were still there from the start. The Abbey Road experience for Converge and the Communitas choir was led and managed from the beginning and a wider collaboration with Converge, although preferable, was not entirely possible (as is written in detail in the following chapter). Collaboration, when it is democratic and treats all contributors equally, takes time, whereas the singular producer taking executive decisions cuts the duration down significantly. It is not surprising that production teams that we have studied (such as Stock Aitken Waterman), have clearly identified decision makers at different parts of the production process. One good example of necessary leadership when it comes to time management is Chris' arrangement of the song for the musicians. They have little or no room for artistic manoeuvre, they merely need accurately (technically and emotionally) to play the parts. The 'house band' for the day are then ready to deliver within a tight timeline of two hours maximum.

Prior to the Wednesday recording session of 'Nothing of Value', Chris's band, Halo Blind, had been recording in Studio 2 on the two previous days as part of his research into production collaboration and the artistic self and to record Halo Blind's next EP. Most of the band were able to stay for the 'Nothing of Value' session on the Wednesday to operate as the 'house band'. This meant (with the addition of Simon Snaize on bass) that they were set up to go straight into recording mode in the morning (9am to 11am). The two lead vocalists, Ed Coulden and Esther Clare-Griffiths, recorded from 11am to 1pm. Communitas, led by Chris Bartram, were in from 1pm to 3pm, leaving time for Jonnie Khan to add brass, Chris to play the Studio 2 Hammond organ and for any additions or retakes. When one is lucky enough to have time in Abbey Road, there is usually little or no time to explore ideas 'in the moment'. On the day, the collaboration is of a different order, it has moved from creativity to execution of performance and recording. At this point the difference between leaders and followers is polarised, it is not time for performers to comment and contribute new ideas. As such, Communitas necessarily came into Studio 2 to take up prearranged chairs with individual headphone sets and were led through the session. At this point they are 'done to' and hence this is where the leaders and producers have to be great with people.

The Studio 2 control room, famously situated upstairs from the live room, was full of musicians, engineers, and producers including Converge's own producer, Faith Benson (see Figure 9.3), one of the coproducers for this track. At the helm as the main recording engineer was Paul Pritchard who, as well as knowing the studio inside out, had the patience of a saint and knew how to handle the people just as well as the equipment, a skill that previous Abbey Road producer, George Martin, rated so highly. As an example of his profound skill at making the choir at ease, for their benefit and for a better performance, Paul played a well-practised trick with the choir, that is, talking directly to them through the headphones whilst not including the conductor, Chris Bartram, who was standing in front of them, in his communication. This 'talking behind his back' approach and the confusion on Chris Bartram's face when the choir started laughing for no apparent reason was magical, an ice-breaking moment. Emotional intelligence, people skills, understanding etc., although rated highly, are less well documented in the field of music production, so there should be room for the unapologetic consideration of small details outside of the more familiar technical realm.

Figure 9.1 The table in the control room of Studio 2

Figure 9.2 Jonnie Khan on trumpet on a Persian rug on parquet flooring

Figure 9.3 From left to right, coproducers Faith Benson, Robert Wilsmore, and Christopher Johnson

Similarly, and of much less importance, on the table by the sofa at the back of the control room, just near the door to the small greenroom, was (the aforementioned) plate of fruit consisting of two pears, three apples (two red, one green), two bunches of grapes (one red, one white) and five bananas (see Figure 9.1).

It would be more familiar ground in discussions on recording sessions to give a technical specification of the equipment used, as in the general equipment and how it is used will have the bigger impact on the recording. But if we flatten the hierarchy we can consider anything that may have causation as an 'actor', as something that has changed something else, and hence worthy of consideration. Not least an actual apple in Abbey Road may act differently to the word 'apple', which is loaded with historical significance, from original sin to The Beatles to mobile phones. Although the returns on consideration of fruit over technology may seem very much smaller, there is a clear advantage in this flattened approach in that it removes the assumptions that established hierarchies and doxa place on events. If we only ever look at the obvious we will miss the other actors in the event, which, when brought together in their legions may (upon return to a hierarchical viewpoint) be more significant than the obvious ones. As part of the environment, as we noted in an earlier chapter, fruit in the control room is indicative of a different approach to that of cigarettes and alcohol in the control room, standing respectively as signifiers of professionalism in the first case and bohemian artistry in the latter.

Coproduction, as understood by Converge, attempts a flattened hierarchy. That is not to say that all things are equal, but that all things and all people, have value. This is what Eno points to in Scenius, that the names of the great artists should encompass all that have acted to create the scene, the event, the genius of the scene (Scenius). Were we to follow a Latourian actor network (ANT) approach and full-heartedly meet the mantra of "describe, describe, describe" (Latour 2007), we would have to observe and describe all actors and actants, so we will have to note that we have privileged fruit here as an example of marginalia and have unfortunately left other actors excluded for the moment. But the purpose is to flatten, to note that things that are not generally considered to hold value actually do so, and it is easy to escalate the signifiers of the habitus that lead to concretised cultural capital and to value; the name Apple rather than the fruit, the black and white aspect of the studio's logo propagating the iconography of the nearby zebra-crossing, the Persian rugs on the parquet flooring (see Figure 9.2), the 'legendary' Neumann U47 and legions of U87s, the remnants of the Echo Chamber to a room off the back not in use as an audio effect but in use as a status effect, the corridors lined with the once-used machinery of recording, and the (reported) seaweed in the acoustic attenuation hanging on the studio walls.

The 'Nothing of Value' project was primarily an *experiential* series of events. The priority was never about record production with regard to making a recording, its primacy lay in the experiences of the participants, and the experiences and aims were multiple. For Converge it was part of realising the five principles for its participants, for music production students it was for their education, for the players and singers (and for everyone) an opportunity to record at this studio, for the academics it was to study a unique environment, for studio staff to earn a living doing the thing they love doing. The song itself remains only as memories and on files, unrealised as a recording. Only a few weeks after the recording session, the UK went into lockdown as Covid-19 became a pandemic. Everyone's attention turned elsewhere and the session files have, to this point, remained unopened. In the end it is the recording that was marginalised, the event and the project with its multiple aims were most valuable. If and when a recording is finally released its main purpose will be, for the most part, as a reminder of the events, just part of the marginalia of the project.

10 Something of Value

Coproducing with Converge, a University-Based Educational Programme for Adults with Mental Health Difficulties (Abbey Road Case Study No.2, Part 2)

Ruth Lambley

The idea of producing a song around the theme of feeling a person to be of value had been around for a number of years, when an opportunity arose at the end of 2018 for the Communitas Choir to record a song in Abbey Road Studios. The Communitas Choir is a community choir and is part of 'Converge', a programme of education for adults with mental health difficulties based at York St John University (referred to as Converge students). In January 2019, the choir travelled to London and recorded the song 'Nothing of Value'. This chapter explores the concept of coproduction in mental health research and then relates this to the creation of the song. I begin with examining how coproduction may be defined and discuss why researchers might want to coproduce their research and identify key features of coproduced research. Based on interviews I conducted with key individuals in the process, I then describe how the song was produced, drawing comparisons between this method of creation and the process of coproduction in research. While coproduction can add considerable time and effort to projects, in mental health there are powerful ethical, moral, democratic, and emotional arguments to justify this extra effort (Beresford 2013). Although it is not a perfect demonstration of how coproduction can be transformative, it was a pragmatic attempt in the context of the very limited time and resources available.

Coproduction in Mental Health Research

How Is Coproduction Defined?

The term 'coproduction' originated in US public management theory during the 1980s, in part in response to the fiscal crisis triggered by the oil shock of the late 1970s (Glynos and Speed 2012). At first 'coproduction' was used to describe the interdependent relationships between citizens and public institutions in resource administration (Lambert and Carr 2018). At this time, movements to challenge professional power and increase citizen participation in community affairs coincided with efforts to reduce public spending (Needham and Carr 2009). In healthcare, it was initially conceptualised by Elinor Ostrom in the late 1970s who described the lack of recognition of service users in service delivery (Realpe and Wallace 2010). Lambert and Carr (2018) cite Edgar Cahn, explaining that his version of coproduction is concerned with societal rather than service transformation – to fundamentally change administration and service delivery, locating power *with the citizen*, rather than just using them to improve the system or service delivery. It is an approach to public services where things are designed and delivered 'with' people rather than 'to' or 'for' them (Horne, Khan, and Corrigan 2013; Nesta 2013). This is part of a much broader paradigm shift in society, a movement towards greater social justice and empowerment. Cahn (2000,

DOI: 10.4324/9781351111959-11

in Realpe and Wallace 2010) explained that "coproduction challenges the assumption that service users are passive recipients of care and recognises their contribution in the successful delivery of a service" (p.10).

The term 'coproduction' has been described as a 'slippery concept' (SCIE 2015) which is used in a multitude of ways and means different things according to the context. An early definition was attempted by Whitaker (1980, cited in Glynos and Speed 2012): "the active involvement of the general public and especially those who are the direct beneficiaries of the service" (p.404). Authors and professionals are frequently unsure of what 'coproduction' means, and its different usage has led some to worry that it has become a 'meaningless catch phrase' (Skills for Care 2018). Coproduction has proved to be difficult to define and pin down, in part because of the range of disciplines which use it, and the considerable gap between abstract definitions and the loose ways in which it is actually applied in reality is perhaps where the 'slipperiness' lies. According to Farr et al. (2020), coproduction is:

> used across different academic disciplines including: economics, political sciences, public administration, voluntary sector studies, public management, science and technology studies, services management, public policy, public engagement and participatory action research.
>
> (p.4)

Oliver, Kothari, and Mays (2019) observed that there are many forms of collaborative research practices, including coproduction, co-design, co-creation, stakeholder and public engagement, participation/involvement, and integrated knowledge transition. According to the Social Care Institute for Excellence (SCIE 2015):

> Coproduction is not just a word. It's not just a concept. It is a meeting of minds coming together to find a shared solution. In practice, it involves people who use services being consulted, included and working together from the start to the end of any project that affects them ... A way of working whereby citizens and decision makers, or people who use services, family carers and service providers work together to create a decision or service which works for them all.
>
> (p.5)

The definition of coproduced research adopted by the National Institute for Health Research (NIHR: Hickey *et al.*, 2018) is of an approach in which:

> researchers, practitioners and the public work together, sharing power and responsibility from the start to the end of the project, including the generation of knowledge. The assumption is that those affected by research are best placed to design and deliver it and have skills and knowledge of equal importance [...] There is no single formula or methods for coproduction.
>
> (p.5)

Hickey *et al.* go on to explain that their approach is principle driven rather than a fixed set of tools or methodologies. As such, it requires that all relationships are valued and nurtured and great effort is used to redress power differentials. Again, the lack of a single formula or methods for how to coproduce in reality could be regarded as 'slippery'. Skills for Care (2018) define coproduction as a process where "professionals and citizens share power, knowledge, skills and experience ... to plan, deliver and monitor services together ...

recognising that all partners have a vital contribution to make" (p.38). According to Farr *et al.* (2020), coproduction builds upon participatory action research and co-operative inquiry, where there is an ethical commitment to challenge social hierarchies, and leads to action that benefits communities, especially those with the least power. Similarly, Slay and Stephens (2013) highlight that a coproduced research relationship is where:

> Professionals and citizens share power to plan and deliver support together, recognising that both partners have vital contributions to make in order to improve quality of life for people and communities,
>
> (p.3)

The basis on which coproducers engage with each other needs to be fair, open, honest, and transparent, and is not about swapping one dominant power base – the professional, with another one – the person with lived experience (Lewis *et al.* 2017).

Needham and Carr (2009) explain how coproduction can be understood on three different levels. The first is 'descriptive', where coproduction is simply a description of how services rely on some productive input from users. This may be just compliance with legal or social norms (like not dropping litter). This approach fails to recognise the potential for a more effective use of productive capacities or of creating new social capital. The second, 'intermediate', is where the contributions of service stakeholders are acknowledged but without necessarily changing fundamental delivery systems or organisational culture. This approach recognises the benefit of user input (which is usually not costed), but the power of existing informal social networks are valued and harnessed, allowing people to become 'more involved and responsible users' who are invited to make a greater contribution to the service. It may include service users being involved in the recruitment and training of professionals. Intermediate coproduction offers a way to acknowledge and support contributions from stakeholders without necessarily changing any fundamental delivery systems. Of concern, this level may involve responsibilities being imposed on service users, perhaps in a manipulative or exploitative manner. The third level is 'transformative' – at its most effective, coproduction leads to the transformation of services. Here, coproduction requires sharing power and control, a change in organisational culture means structures of delivery should entrench coproduction, mutual trust, and reciprocity between professionals and communities. New relationships are forged, service users are positioned as experts, and there is a 'whole life focus' (Needham and Carr 2009, p.9) looking beyond, to clinical or service issues, to recognising a broader quality of life. Service users are involved in shaping the organisational ethos, and both service users and frontline staff are empowered. Needham and Carr cite Bovaird (2007):

> The service user has to trust professional advice and support, but the professional has to be prepared to trust the decisions and behaviours of service users and the communities in which they live rather than dictate them.
>
> (p.6)

Transformative coproduction demands and is dependent on a fundamental paradigm shift towards legitimising and valuing experiential and first-hand knowledge (Lambert and Carr 2018). Needham and Carr (2009) observe that the challenges and dilemmas of coproduction depend very much on the type of service being coproduced. However, while some service users are ready and able to participate in the opportunities arising from transformative coproduction, others may remain socially and personally disadvantaged – this situation needs to be carefully considered and managed (Needham and Carr 2009).

Table 10.1 Arnstein's ladder of degrees of citizen participation

8	Citizen control	Citizen control
7	Delegation	
6	Partnership	
5	Placation	Tokenism
4	Consultation	
3	Informing	
2	Therapy	Nonparticipation
1	Manipulation	

Slay and Stephens (2013) cite Arnstein's (1969) ladder of participation (Table 10.1). The ladder shows the movement from 'non-participation' to citizen control. At the bottom of the ladder is non-participation – traditional services at their most coercive, for example, applying manipulation and therapy without input from citizens. In the middle is 'tokenism', which can be seen as inviting some shallow engagement but where parameters are set by the professionals. At the top of the ladder is 'citizen control' where there is a fundamental shift in power to the service users, with the creation of equal and reciprocal relationships and shared decision-making power.

Key Principles and Features of Coproduced Research

Several authors have listed key principles of coproduced research. Table 10.2 shows the principles listed by Slay and Stephens (2013), Skills for Care (2018), NIHR (Hickey 2018), and Farr (2020). There are clear similarities in terms of reciprocal relationships, sharing power, relationship building, and recognising and utilising the skills of all. It is important to establish ground rules, creating an environment where all voices are heard and treated with respect. There should be on-going dialogue throughout the project, including joint ownership of key decisions. There should be a commitment to relationship building that addresses power differences, and creating open, honest, trusting, and reciprocal relationships. Safe spaces should be created. There should be opportunities for personal growth and development, including appropriate training, with an emphasis on support, treating people as assets and unlocking their potential. Coproduced projects should be flexible, creating an iterative, fluid, open-ended, experimental and interactive process. Lastly, there should be continuous reflection, where team members reflect on how they are working together and using their expertise and perspectives on the project.

The following section explores the creation of the song 'Nothing of Value' in the context of this theoretical background. I will draw comparisons between the process of coproducing research described here, and the process of coproduction that occurred in the development of this song.

The Creation of 'Nothing of Value'

The 'Communitas Choir' is a community choir and part of Converge, a programme at York St John University that provides educational opportunities to adults (hereafter termed Converge students) with mental health problems. The choir has over 30 singers, drawn from Converge students, university students, and university staff. Universities can often be considered as fairly hierarchical institutions; however, parts of York St John University can be considered a fertile ground for trying new and creative things, demonstrated, for example, by the existence of Converge. The idea of creating a musical project around the idea of

Table 10.2 Principles of coproduction

	Slay and Stephens (2013)	*Nesta (2013)*, **Skills for Care (2018)**	**NIHR (***Hickey et al. 2018; Farr et al. 2020***)**
Assets	**Taking an assets-based approach:** transforming people from passive recipients to equal partners	**Assets:** working with people with lived experience as equal partners, transforming the perception of people from passive recipients/burden on the system	**Building on people's assets:** valuing skills, knowledge and capabilities (Farr et al. 2020)
Capabilities	**Building on people's existing capabilities:** recognise capability and support the person to use them	**Capacity:** recognising and growing people's capabilities and actively support them to put them to use	**Including all perspectives and skills:** Making sure the team includes all those who can contribute, involving diverse stakeholders (Hickey et al. 2018) **Respecting and valuing the knowledge of all:** everyone is of equal importance (Hickey et al. 2018)
Reciprocity/ mutuality	**Reciprocity and mutuality:** range of incentives to work in reciprocal relationships with professionals, mutual responsibilities and expectations	**Mutuality:** Working in reciprocal relationships where there are mutual responsibilities and expectations	**Reciprocity and mutuality:** Everybody benefits from working together (Hickey et al. 2018)
Networking	**Peer support networks:** engage as a way of transferring knowledge	**Networks:** Engaging peer and personal networks alongside professionals as the best way to share knowledge	**Building and maintaining relationships and shared learning:** an emphasis on relationships is key to sharing power, safe spaces should be created (Hickey et al. 2018)
Blurring distinctions	**Blurring distinctions:** removing the distinction between professionals and recipients, reconfigure how services are developed and delivered	**Blurred roles:** Removing tightly defined boundaries between professionals and people with lived experience, changing the way in which services are designed and delivered	**Sharing power:** The research is jointly owned and people work together to achieve a joint understanding (Hickey et al. 2018) **Blurring roles:** reducing boundaries between professionals and service users (Farr et al. 2020)
Facilitating rather than delivering	**Facilitating rather than delivering:** ensuring agencies become catalysts and facilitators	**Catalysts:** engaging public service agencies to become facilitators rather than central providers	
Creativity			**Being innovative and creative** with room for experimentation (Farr, 2020)

feeling like someone of value involving the Communitas Choir had existed for a number of years. When the opportunity to record a song at Abbey Road studios arose, the decision was made to include Converge students, both choir members and song writing students, in the production of the song. Coproduction was sought here due to the view that it is of intrinsic value – fairer, more ethical, and more democratic, and also that it might have more impact in terms of interest generated by the project. Ideally, everyone would have preferred a slower and more collaborative lead into the project, but the deadline of the session at Abbey Road meant that the song was perhaps not as coproduced as those involved would have liked.

As coproduction in mental health research has different levels and meanings, collaboration or coproduction in music has different approaches. This ranges from a top-down approach where there is a leader who steers everything, to a complementary approach where people do the part that corresponds to their skill, to an integrated model where everybody does everything and nobody really takes charge (see John-Steiner 2000). The production of this song falls into a complementary approach.

Why Would You Want to Coproduce This Song?

In terms of the four arguments for coproduction expressed by Oliver *et al.* (2019), first the 'substantive' reason would be that coproduction might improve the quality of the music, in particular, its relevance to members of the choir. The second 'instrumental' reason would be that because of being coproduced with Converge students, it might have more impact, for example, in terms of the interest generated in the song, and the extent of the dissemination of the music. The third reason, 'normative' is perhaps the strongest justification in the context of producing this song, that coproducing alongside Converge students is of intrinsic value – being fairer, more ethical, democratic, and enjoyable.

What Would It Look Like to Coproduce a Song in the Mental Health Sense?

As described previously, Needham and Carr (2009) explain how coproduction can be understood on three different levels: descriptive, intermediate, and transformative. Transformative coproduction is the most effective level of coproduction. It requires a relocation of power and control, and development of new mechanisms of planning, delivery, management, and governance. Table 10.2 shows the principles of coproduction as suggested by several authors (Slay and Stephens 2013; Nesta 2013; Skills for Care 2018; Hickey *et al.* 2018; Farr *et al.* 2020). What would it look like to apply these principles to the production of a song with a community choir? Here, an assets-based approach would see all those involved in the creation of the song as equal partners, where the skills, experience, and knowledge of all participants are valued. Individuals' capabilities, for example, their lyric or melody writing skills, will be recognised, supported, and utilised. A commitment to relationship building would address power differences, and create open, honest, trusting, and reciprocal relationships between those involved. For example, the creator of the instrumentation, the band, the song writers, the soloists, and choir, would all benefit from working together. There would be joint ownership of key decisions, for example, the final lyrics and the feel of the instrumentation. The project would be jointly owned, and traditional boundaries and distinctions between musical professionals/experts/specialists and people with more lived experience would be blurred. There would be opportunities for personal growth and development, training, support, treating people as assets and unlocking potential for all participants. Finally, the process would catalyse the different agents involved to become facilitators of producing the music, in innovative, experimental, and creative ways.

As observed by Locock *et al.* (2016) universities are inherently hierarchical – in a truly transformative process of coproduction there would be no hierarchy. The inspiration for the song would emerge collaboratively from the people who would sing it, compose the music, and direct it, rather than coming from a professional. The lyric writing process would be a collaboration between the singers and songwriting specialists, and the process of creating the melody and instrumentation would be shared between the expert musicians and the singers. This process would take time, to allow all voices to be heard, especially those who may struggle to express themselves perhaps due to mental health needs. The different types of expertise among participants (professional or otherwise) would be recognised and utilised. Power processes would be considered; for example, is there a process of 'gate keeping' where professionals select who can participate such as choosing who can sing in the choir or write the lyrics? Are the sessions being conducted in an accessible way in terms of time, location, and format? Is the input of participants being shaped or manipulated by those with more power, for example the process of writing lyrics or creating the instrumentation?

Farr *et al.* (2020) give an overview of things that should be considered in a coproductive process. In particular, they observe that coproduction can add considerable time, effort, and expense to a project. An important aspect in the context of creating a song is their point that people should take on roles that suit their personal strengths and skills, which means that coproduction is not necessary at each stage – it is not a case of everyone doing everything. This might mean, for example, that a professional musician could lead on the creation of the instrumentation, perhaps alongside someone with less expertise but is keen to learn – opportunities for training and development are an important component of coproduction. Due to this, there should be an ongoing conversation between all parties about the extent to which they wish to participate in which components – are they happy just singing, do they want to be involved in lyric writing, or learn about the process of creating the instrumentation? Decision-making should be distributed among all those involved; in particular, it should be agreed which decisions and aspects of the project really need to be coproduced, for example, can the professional musician create the melody without input from all those involved? Do the lyrics need to be agreed by the choir? A particular consideration when working alongside people with mental health difficulties is that this may cause them to fluctuate in the extent to which they are able/wish to participate.

The Coproduction of 'Nothing of Value'

Coproduced projects should be flexible, creating an imaginative, iterative, fluid, open-ended, experimental, and interactive process. Skills for Care (2018) identified factors that might facilitate coproduction. First, 'fertile ground' where there is an existing culture of involvement and inspirational leaders who are prepared to promote coproduction. Second, there need to be structures and processes to sustain coproduction, for example, training resources, relationship building skills, and flexibility to make reasonable adjustments according to a person's needs. Finally, they highlight the need for a 'give it a go' culture, where they persevere even if they can't achieve the fullest coproduction at the beginning. York St John University can be considered a fertile ground for trying new and creative things, with inspirational leaders prepared to take risks, and a strong belief in emancipation and ethics. This is clearly demonstrated by the creation of Converge and the new Institute for Social Justice.

Despite the slow progression from the initial idea (which lay dormant for several years), the production of the song happened quite quickly when an opportunity arose at the end of 2019 for studio time at Abbey Road Studios in London. The project began in November with a studio date at the end of January 2020, meaning there was a lot of work going on over the Christmas holidays. Most of the communication was over email. Figure 10.1

shows the process of producing the song. Robert Wilsmore had written the chorus and approached Nick Rowe, the Director of Converge, Esther Griffiths, who is a songwriting tutor at Converge, and Chris Bartram, the choir's director, to develop the song. Chris Johnson from the Music Production department had written a melody for the verses and the middle 8, and the instrumentation. A demo was made in the studio at York St John to take to the choir.

Robert met with the choir to explain about the background of the project and to explain that the choir would sing the chorus of the song. The demo created by Chris Johnson was played, and Esther explained how after the choir had sung through the chorus a few times, which includes the lyric "Are We Nothing of Value?" they:

> all felt that by the end of it we wanted it to be a bit more positive, because we were all starting to feel it was … making us feel a little bit down, so I brought it to him [Rob] that maybe we could just end it on a slightly more upbeat note … He said yes … so we ended it with "We are something of value to you!" The main chorus lyric changed to a more positive end.

Robert recalls the same situation:

> even at the [choir] rehearsals we were changing things. Them [the choir] coming back to us and going "yeah can we not do that, can we do this instead?" Including changing the words at the end, you know, we changed the words from "Are we nothing of value to you?" to "We are something of value to you!" And that came from the group saying "no, no, this needs to be a really positive statement, not a question, but a positive, proactive statement", and that was really lovely.

Chris Bartram identified that:

> [Rob and his team] had a clear idea what they wanted in terms of sound and 'feel', but there was scope for me to develop this with the choir, to create the energy, and to work on this so that it still felt 'like Communitas'.

Esther was able to involve the Converge songwriting students she teaches as a Converge course – most of whom are also members of the choir. Together they worked on the lyrics of the verses to get something structured and create a rough demo.

Prior to the studio date, the choir had been rehearsing the song for about a month. Due to the studio size in London and the expense of the trip, it wasn't possible for all choir members to travel, so it was decided those who had been members longest would be offered the opportunity first.

On the Day …

The choir caught a train from York to London at 9.30 am, with one member saying they had stayed up all night so as not to miss the trip! In the morning before the choir arrived, Rob and some students from the Music Production course helped set things up at the studio. Chris Johnson had arranged the song for his band Halo Blind, and in the morning, they recorded the track – drums, guitar, and bass, so the choir and soloists would have something to sing along to in the afternoon.

After a trip across London where, as one student described "the tube tunnels seemed endless", the choir arrived at Abbey Road Studios. A student recalled how:

> When we arrived at Abbey Road everyone stood on the steps for a photo and also before we left, with Chris Bartram leading us in singing the Beatles song "Help", much to the amusement of passers-by!

Faith Benson, producer for Converge and coproducer on for 'Nothing of Value' took the choir to the canteen where they ate their lunch and took in the atmosphere. With the slight delay to the schedule, Faith tried to keep people as relaxed as possible. One student recalled how:

> On the walls there were photographs of famous faces – the Beatles, Ed Sheeran, etc. who had recorded there. It was wonderful to be there, so exciting, something that none of us had ever dreamed would be possible.

When the time came to record the song, Faith ensured that the engineers knew the choir had never sung with headphones on before:

> singing with headphones on is quite different than when you are singing without them because you aren't used to hearing your own voice, [so I told the engineers] that's something we might have to look out for, how well [the choir] deal with it, and what we can do if they don't deal with that well. But they all did really well with that, so that wasn't an issue.

Chris Bartram recalled his role:

> it was up to me to warm up the choir, get them through the initial 'wow factor' of being in Studio 2 and focus them in, and support them to develop the energy and sound so that they gave everything in the recording.

Esther remembers the process:

> We had a little warm up, and … just went for it really, did a few takes and managed to get it really quickly done … And then I did some of the solo vocals and another guy, Ed, … did some of the other solo. And … all of us just kind of teamed together and did some harmonies as well.

A choir member recalls:

> We recorded the choruses in about half an hour … this went smoothly and a couple of times the people in the control room … came down to ask us to sing it slightly differently but it was soon completed and we were able to take photographs in the studio.

The Experience of Going to London

Aside from the singing, for many choir members, the trip to London was an immense experience. Some had never been to London before, and for others, going to such an iconic studio gave them a huge sense of pride.

Esther commented:

> Everybody absolutely loved going down to London and the experience of being at Abbey Road was really good fun … we had lots of photos taken on the zebra crossing … and

> outside the steps … The feedback that I got from the students was really positive … they just really enjoyed the experience … it was, I think it was a very great chance for a bit of camaraderie on the train back … singing Beatles songs and things.

Nick also recalled:

> It was fantastic, wonderful for the choir, they still remember it. It was a very special day. I mean a lot of people, some of the people had never been to London before … so it was an important event for the choir, you know, and in the way that they thought of themselves, and crossing over the famous zebra crossing outside and all of that kind of thing, so it was a really important event in the life of the choir.

A choir member remembers:

> I feel very lucky to have been part of this journey, as I have been a fan of music such as the Beatles who recorded there. It was not something that I ever dreamed I would be able to do. We felt very valued by being chosen, and it was so nice to be with friends I had made in the choir. All the leaders were great, very supportive and kind … We had a lovely day, something that I will never forget. I would love to hear the finished song, now all delayed due to lockdown. Thank you so much Communitas Choir and York St John Converge for making this possible.

Reflections on Coproduction in This Project

In Figure 10.1 we can see the project hierarchy of power in the production of 'Nothing of Value'. In a typical situation, at the top of the hierarchy is the owner of the project, Robert Wilsmore, who conceived the original idea for the song. It was he who recruited the Director of Converge, the choir's Director, Esther the songwriting tutor, and Chris Johnson to write the melody for the verses and compose the instrumentation. At the bottom of a traditional hierarchy lies the Communitas Choir, the soloists, and the band, Halo Blind, who sung the lyrics and played the music produced by those above them. Whilst in a truly transformative coproduction there would be no hierarchy, this project, however, could arguably be seen as an imperfect and pragmatic attempt to coproduce within the given time and resources. The willingness to try and coproduce from those with power was there, and there are aspects of the process of creating this song that are transformative. For example, if Converge students are seen as the group with which the musical experts coproduced with, alongside Esther, Converge songwriting students were able to shape the lyrics sung by the soloists, and the Communitas Choir's request to change the lyrics of the chorus was embraced by all participants.

Esther reflects on the coproduction in this project in a very musical (rather than research) sense, concerned with provenance rather than emancipation and social justice:

> [it's] working together on a project and coming up with ideas together and perhaps, just sort of brainstorming different ideas, different approaches, and then kind of choosing together … within music, that is really exciting, because you can have one person coming up with the melody for the verse, and some other people coming up with lyrics that that they want to express… it's a very wide spectrum of different people on board with it as well, which was really, really interesting, from music lecturers to choir people, to me, to the guy who was the vocalist … a range of people all working towards the same goal …

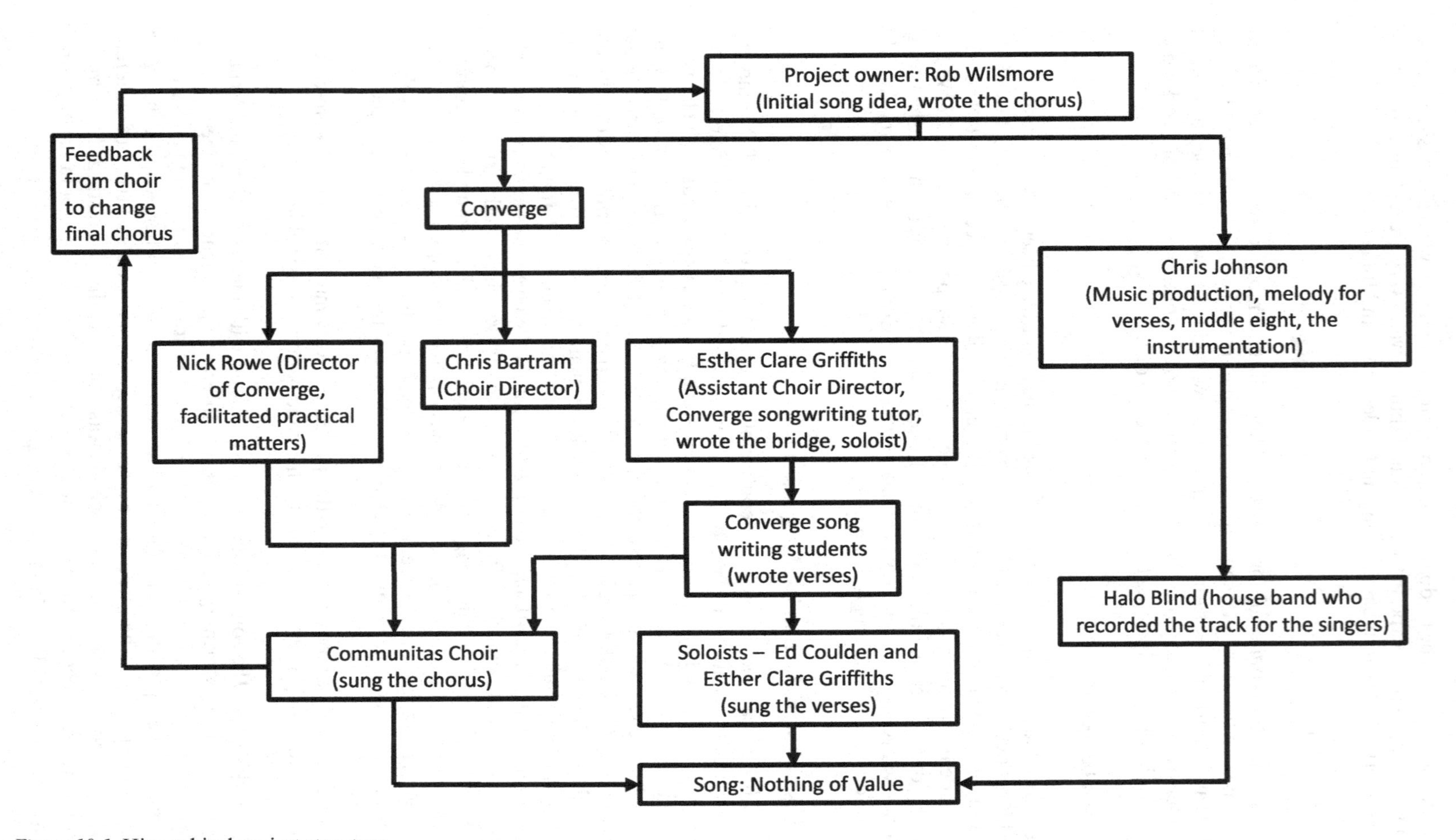

Figure 10.1 Hierarchical project structure

Both the director of Converge and the choir's director acknowledged the shortcomings of coproduction in this project – the song had largely been determined before the song was brought to the choir:

> [Nick] One could argue that a truly coproduced event would involve the choir from the very beginning in what the song was going to be about, the words [and] the music.
>
> [Chris Bartram] For the Abbey Road project, we had little say in creating the song itself or the final product, as this had largely been determined and worked out beforehand: Rob and his musical team were fairly clear what they were wanting Communitas to do in singing the chorus, and the way they wanted it delivered.

In reality coproduction, in terms of contribution to the song, had been musical and lyrical input from Esther as a Converge leader, and with lyrics from Converge members, some, if not all, of whom were also part of the Communitas Choir. So, there is a slight nuance in that Communitas became the performing body, whilst Converge became the creative contributing body.

Looking at coproduction in a slightly different way, Nick goes on to highlight the coming together of people who do not usually work together:

> a group of people with lived experience, together with the Abbey Road studios, together with lecturers and staff at York St John … That coming together was innovative and coproductive.

Similarly, Chris Bartram observes coproduction in a different way:

> As I write, I'm thinking that 'coproduction' in the largest sense also involves the role of myself and the Communitas staff team (the choir's Assistant Director Esther Griffiths, student co-leader, other music students, Nick and Hazel Rowe, etc.) in preparing, organising and getting the choir down to London, across the busy city to Abbey Road, and back, all in one day. There was a lot of liaison and logistical planning involved in making sure this was achieved successfully and smoothly, in order to ensure that the choir were in the right positive physical and mental place to make their contribution effectively in the studio. This I also understand to be an important element of the coproduction.

Conclusion

This chapter explored the process of coproduction in mental health research, and compared it to the production of a song, 'Nothing of Value'. Although it is not a perfect demonstration of how coproduction can be transformative, it was a pragmatic attempt in the context of the very limited time and resources available. Truly transformative coproduction can add considerable time, effort, and expense to a project. In addition, while universities can often be considered as fairly hierarchical institutions, those involved in this project felt that the coproduction improved the final product, and believed that the efforts made to coproduce was justified in terms of being a fairer and more ethical process, and also that coproduction might increase the impact of the song in terms of interest generated and in its dissemination.

Part Two

Type 2. Internal Coproduction

The Self as Many

11 The Artistic Self and the Cycle of Production

Christopher Johnson

This chapter considers how the creative persona of a producer–artist is affected in the practice of writing and producing songs. It examines the stages of music production, suggesting that they are better understood in cyclic rather than linear models. It investigates internalised forms of collaboration and uses of technology in contemporary song production and how they affect the producer–artist's development of their creative Self.

The tools, expectations and social contexts of modern music production have become embedded in compositional and musical arrangement practice. The Digital Audio Workstation (DAW) is at the centre of most production processes, and its accessibility has allowed a growing number of artists to produce themselves. This DIY producer is often a sole creative able to and required to fulfil multiple creative, technical, and entrepreneurial roles within a production, from the initial musical idea to the final mix, and as such the production journey presents countless opportunities for them to express an artistic persona in the sonic worlds they create. They can distil elements of their visual image, personal history, musical influences, and world view into a curated artistic Self that is represented in their production sound. In Burgess' typology of music producers these artists can become auteurs defined by "the unique personal identity with which they infuse their productions" (2013, p.10). However, we must be careful not to reinforce a romantic ideal of the producer–artist working in isolation, crafting their glorious sonic art in moments of fevered inspiration in communion with their muse, or even just their DAW. The critiques of such romantic mythology around record production are examined by McIntyre (2012) who sees creativity more as "the emergent property of a system in action" (2019, p.69) and demonstrates comprehensively across numerous writings that creativity in popular song production is better understood as the result of collaboration, even if the collaborators are unseen and unmet actors in a wider system that the producer is working within (McIntyre 2019, 2012, 2008, 2011).

One perspective on how such distributed collaboration could occur is presented by the pioneering Russian filmmaker Sergei Eisenstein. He identified the consciousness of the spectator as the site where meaning is created but discusses this creativity as a collaboration with the text's writer in which the writer attempts to guide the spectator's response, and in which the spectator's contribution is dependent on their unique history and disposition:

> In fact, every spectator, in correspondence with his individuality, in his own way and out of his own experience – out of the womb of his fantasy, out of the warp and weft of his associations, all conditioned by the premises of his character, habits and appurtenances, creates an image in accordance with the representational guidance suggested by the author, leading him to understanding and experience of the author's theme. This is the same image that was planned and created by the author, but this image is at the same time created also by the spectator himself.
>
> (Eisenstein 1947, quoted in Gruber and Davis 1988, p.261)

DOI: 10.4324/9781351111959-13

Barthes places more emphasis on the reader in his discussion of 'The Death of the Author' (1977) and reduces the creativity that can be credited to the author. Barthes' scripter, thereby, combines many previously existing elements from a cultural store into a 'tissue of quotations' (1977, p.146) rather than authoring a unique text from scratch. Like Eisenstein, Barthes identifies the mind of the reader as the real site where creativity occurs and meaning is assembled, as it is the space that "holds together in a single field all the traces by which the written text is constituted" (Barthes 1977, p.148).

This perspective would suggest our producer–artist has little agency at all in the creation of their works, as noted by McIntyre (2009, 2012) who instead sees the individual producer–artist as an active part of cyclical system, drawing from and contributing to a dynamic cultural and social sphere with each new creative contribution to it. Drawing on confluence approaches based in psychology and sociology he cites Hennessey in asserting:

> Creativity cannot be separated from the societal and cultural contexts in which it arises [and] only with the adoption of a truly integrated systems perspective can researchers hope to ever understand the complexities of the creative process.
>
> (Hennessey 2017, in McIntyre 2019, p.69)

McIntyre adopts Csikszentmihalyi's systems model of creativity to illustrate this. An individual acquires the body of knowledge of a cultural grouping, or 'domain', that they are immersed in, learning its particular rules, expectations, habits and requirements. This individual is then able to use this ingrained knowledge to create something that offers novelty to the domain. A social group, the 'field', holds the body of knowledge of the domain and assesses the individual's creative contribution. If it is accepted then it enters the domain, which is now slightly altered by the new addition (Csikszentmihalyi 2013, pp.28–29). Subsequent contributions to the domain will be prepared by individuals whose ingrained knowledge would have been modified by the new addition, and so the cycle continues with the domain mutating and evolving with each new contribution in a Darwinian progression of cultural development. Applying the systems model to the discipline of music production, we could say for illustration that the cultural domain is a particular musical genre that represents the entire body of produced works in that genre, say, prog rock. The field is then the fans, music press, and communities that hold the body of knowledge of prog rock and comment on it, and the individual is an artist who, having immersed themself in prog rock and learnt its structures, rules, and production traits to the point where they come naturally, produces an original prog rock song that follows the rules enough to qualify for admission into the prog rock domain whilst simultaneously offering something novel. If the fans, prog rock communities, and music press accept this new song with its novel addition it becomes part of the prog rock domain, contributing to the overall evolving body of prog rock productions.

If we rewind to the conception of the song, and consider the circumstances that affect its composition, then we can examine how the producer–songwriter begins to construct and curate an artistic persona that will be presented to the domain through the song. In band collaborations, this constructed Self must navigate the production process and survive the addition of the other band members artistic personalities who now share the produced audio artefact as a vehicle on which it will be presented to the audience. To endure intact the producer–songwriter's persona must combine successfully with the others to present an enhanced version that will be accepted by the intended audience. This would be the band's artistic persona, a recognisable creative and commercial entity.

However, with the establishing of the DAW as a songwriting tool (Bennett 2018; Strachan 2017; Marrington 2011; 2016; Miranda 2012), the roles of producer, songwriter and musician have become blurred, and the producer–songwriter can now work in isolation to produce the meaning of a track before presenting it to the rest of the band (or indeed their own internalised collaborators). As well as shaping their own creative persona they are shaping the meaning of the song in complex layers of sound that are far more expressive than lyrics, chords and melody alone. Indeed, as indicated by Zak (2001, p.27), the 'song' may not exist until various signifying timbres, musical parts, spatial effects, and other elements not traditionally considered as 'songwriting' are present.

On Persona

At the start of his study on the meaning of popular song(s), Allan Moore suggests that for many of us, music and what it means to us personally will be a factor in defining who we are (2012, p.1). He is concerned with how pop song productions are communicating meaning to the listener. What factors in the aural landscape they present are conveying messages to the listener, from traditional musical domains such as melody, form, and harmony that are inherent in the songwriting, to factors more attributed to the manner of their production, such as timbre, space, and vocal tone? In his book, Moore addresses these questions from an analytical position as listener and musicologist and offers a wealth of detailed case studies from the past century of popular song(s).

In analysing how listeners connect to a song and identifying what the main conduit of communication might be, Moore suggests that it is primarily through the persona of the vocalist, "an artificial construction that may, or may not, be identical with the person(ality) of the singer" (2012, p.179). This persona must in some way become known to the listener so that they can interact with it in experiencing the song. The vocal, as it exists within the track, must carry sufficient information to allow us "to construct the persona of the individual to whom we are listening, on the basis of the track we are listening to" (Moore 2012, p.180). In analysing this embedded information, the semiotic coding of the recorded voice is explored by Frith who points out that we don't just hear recorded voices as isolated phenomena, but "we assign them bodies, we imagine their physical production" (1996, p.196). The recorded voice then, becomes an index of gender, age, ethnicity, class, as well as attitude and character. These attributes form a personality in the understanding of the listener that they can then interact with. This singer–personality that the listener constructs is also affected by factors outside of the recorded track that the listener may or may not already know about the singer in question, from images of their physical appearance and visual stylings in the artwork or online, to how they come across onstage, in interviews or in other media stories associated with them. Moore terms this composite real-world identity the *performer* and discusses it as distinct from the *persona*, which is the voice the performer assumes when singing (2012, p.181). He examines the relationship between the persona embodied in the vocal and its environment (the accompaniment) and focuses on the meaning that can be derived from the ways this relationship is handled, not just in the parts as they are written musically, but in the complexity, timbres and spatial characteristics of their sound worlds (2012, p.187). This would suggest that there is a quality intrinsic to the produced song that communicates a unique character that is a result of the vocal persona and the manner in which the accompaniment is arranged around it. At a simplistic level, we could say that crafting this *production character*, or the track's personality, is one of the core roles of the producer, and that overseeing the manner of the contributions of various collaborators on a project and handling all the required logistical, technical, musical, and social arrangements should be carried out in service of this craft. In illustration, Zak examines the sonic distinctiveness of particular instrumental, spatial, and production timbres unique to the situation in which a track was produced. He proposes that a recorded song contains three compositional layers: the song as it could be transcribed on a lead sheet; the arrangement as the written instrumental parts; and the track, which is the sound recording itself, a set of "specific performances and sounds, of fixed timbral and sonic relationships that the listener experiences" (2001, pp.24–25). Zak's position that the track as a sonic artefact is 'composed' in a creative process related to, but distinct from, the composition of the song and the arrangement, brilliantly draws attention to the concept of the track as a text that is authored in the collaborative production process:

> Compositional decisions are based on or influenced by a similar set of factors, which include technical competence, aesthetic belief, the needs of personal expression, a sense of style and language, a feeling for form and dramatic narrative, familiarity with other works, and so on.
>
> (Zak 2001, p.38)

The song and the arrangement are encountered through the track, which has its own deliberately crafted identity or personality. He goes on to examine well-known individual tracks and identifies examples of 'sonic character' that define a track's personality, such as the 12-string guitar tone of the Byrds' version of 'Mr Tambourine Man' (1965), pointing out that such "striking sonic character adds a further dimension of sensory meaning to the pitch and rhythmic elements of the musical part, creating a comprehensive formulation of musical substance" (2001, p.60). In our consideration of persona and how it is expressed in an audio artefact created by the producer–artist, we can use Zak's three compositional layers as a framework. That is, we can ask how is the artist–producer's persona expressed in the songwriting; how is it handled through the arrangement; and how is it present in the committed performances, timbres, and soundscape that represent the definitive, produced work?

Zak's work examines the aural production character of produced popular songs and how particular recorded performances and sounds can gain a layer of meaning as triggers of association tied to a particular artist, producer, or genre, and Moore's work investigates what this meaning could be and, more importantly, *how* the sounds are meaning what they mean. This all provides a useful starting point for an exploration of songwriting and track production that considers the meanings and character communicated in a recorded song as intended by the producer–artist working alone, and the factors involved in artistic persona and its links with track personality. In this chapter and the next we will be concerned with the persona of the producer–artist and how it develops; the themes of the project; the influence of different internal collaborators; the character of the production process; and how it all comes across in the final product.

The Stages of Production

In considering the emergence of an artistic persona in music production, and how it sits in the character of the production culture or 'habitus', it is important to examine the creative, logistical and technical processes of production, and how they cluster into activities that progressively develop the musical material. In production musicology these are often described in a staged linear process (e.g., McIntyre 2012; Hepworth-Sawyer and Golding 2011), starting with the composition of the material and leading through its development and the logistical organisation of the project into pre-production, then to the committing of performances and timbres in the production stage, and then on to the balancing, sweetening and preparing of the material for market in post-production. It is a staged model familiar to many forms of media production.

It can be difficult to describe to the layperson just what it is that music producers do, and the pre-production – production – post-production linear model can be useful to help explain our activities, or even to help the freelance producer in costing their services to clients. However, it has become a less relevant logistical model for the modern DIY producer, as the creation of their music rarely follows such a staged linear workflow, and so it is not a useful analytical model for scrutinising the creativity of writing records in any depth. The emergence of the DAW as a songwriting tool is a much-discussed factor in the blurring of boundaries between the stages of production (Bennett 2018; Strachan 2017; Hepworth-Sawyer and Golding 2011; Miranda 2012; Marrington 2011; Marrington 2016). For many rock and pop producers the composing of a track is now such an amorphous process that the boundaries between the stages have lost logistical significance. A producer may well be performing tasks traditionally grouped into the mixing post-production stage long before the song has an established structure or harmonic sequence. Similarly, creative tasks that are traditionally considered pre-production are now commonly delayed until near the end of the production process. For example, the lyricist may need to hear the complex interaction of the musicians' performances, and the personality and emotion they portray when carefully balanced against each other in the mixing process, before they can draw out the specific meaning of the song and write the right lyrics. As Zak (2001) has discussed, the very impetus to write the melody and lyrics to create 'the song' from the layers of musical material may only be generated after the producer–songwriter has been inspired by a blending of rhythms and timbres achieved by applying particular effects and mixing techniques more traditionally associated with post-production.

In studying collaborative songwriting environments Bennett (2011) contends that six processes are involved in developing musical ideas – stimulus, approval, adaptation, negotiation, veto, and consensus. A stimulus idea is presented by one writer, which is then either approved, vetoed, or adapted by the other. If both agree the idea achieves consensus and is included in the song as it develops. If vetoed the idea may be negotiated or further adapted. I suggest an adaptation to Bennett's stimulus-evaluation framework to deal with producer–artists working on their own.

The DAW-based track-writing engine (Figure 11.1) is a similar process of experimentation, assessment and rejection that is driven by stimuli. A stimulus is identified following a period of play, usually on an instrument or time spent messing around in a DAW. Countless other ideas have been instantly rejected until one is judged to have something special about it. This idea is developed through improvisation and further play until a point where it is captured and saved so it can be remembered. This kick-starts the engine and the entire track-writing process. The captured idea is usually assessed, refined again, and recaptured, all with a sense of excitement that the idea has merit. This developed idea now becomes the stimulus for the next idea as the artist assesses what the track needs now, for example another instrumental part to harmonise that last one, a lyric idea, or a particular timbre or feel that can be achieved through processing and effects. The new idea then goes through the same process of play, assessment, rejection, or development until it becomes the stimulus for another idea. In this way, subsequent stimuli are self-generated by the engine and on a good day the artist can achieve flow, the state of being completely involved in an activity that is challenging but intrinsically rewarding: "You're right in the work, you lose your sense of time, you're completely enraptured, you're completely caught up in what you are doing" (Csikszentmihalyi 1996, p.121). One idea stimulates the next and the cycle continues with

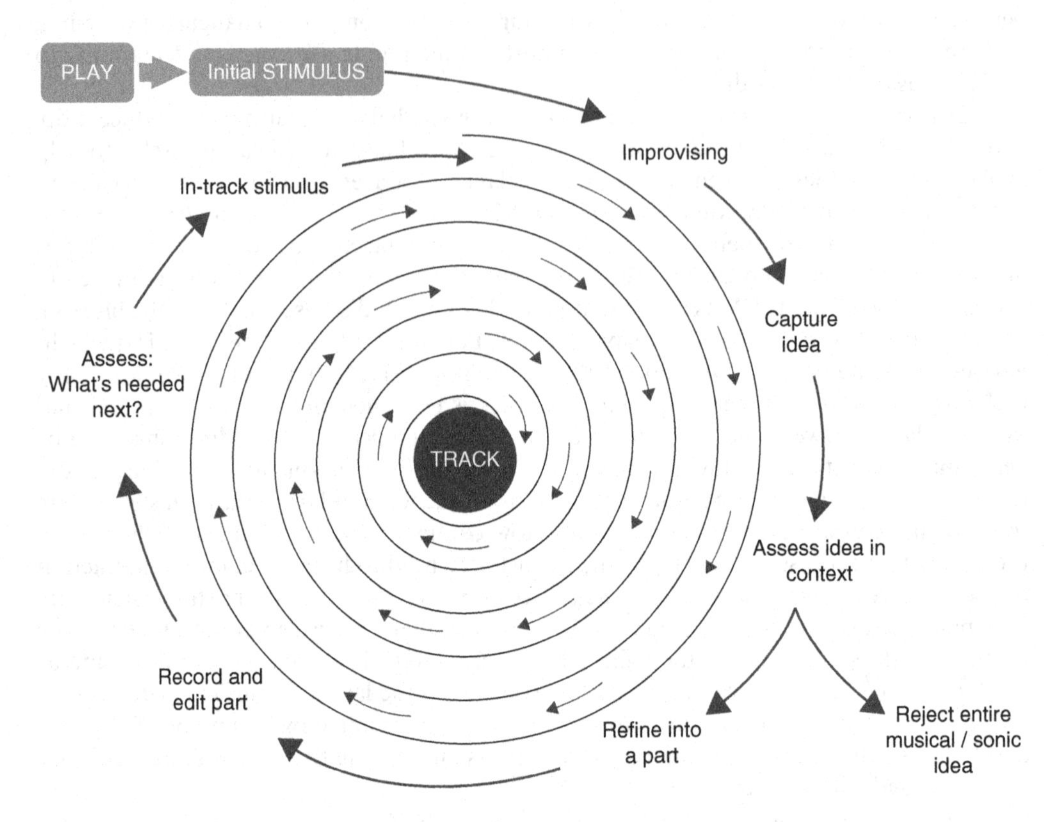

Figure 11.1 The DAW-based track-writing engine

ideas being assessed, rejected, or refined until a complete track is created. Working in the DAW in this way, any process associated with pre-production, production, and post-production is up for grabs at any moment of the track creation. The track develops horizontally on the timeline in terms of its song form structure, but it simultaneously develops vertically on the timeline, where the rhythms, arrangement, density, and timbre are crafted depending on what the current stimulus inspires or demands.

It is worth noting here how these moments of inspiration are the driver of the track-writing process, but we need to examine more closely how the producer–artist decides which ideas are good and why they have value. Gruber and Davis propose a constructionist evolving systems model and use the metaphor of a lightning strike in a thunderstorm to illustrate how sudden moments of creative insight are better understood as part of a complex system in which knowledge, skills, and dispositions build up in the creative individual over time, progressing and interacting and at times producing a flash of insight (1988, pp.243–244). If we adhere to Csikszentmihalyi's systems model perspective to scrutinise this 'inspiration' we might see that it happens when the producer–songwriter hears something in their work-in-progress that they judge is novel and exciting whilst remaining aesthetically acceptable to the field. This moment of recognition where the producer–artist identifies a musical idea or sonic nugget that excites them is critical, both in providing a stimulus to drive the creative process – that crucial intrinsic motivation – and in forming key components in the sound world that determine the product's character and carry messages about the artist's persona. From their position as an artist steeped in the bodies of knowledge of multiple intersecting discipline-specific fields and domains, their judgement of what is novel and exciting will be largely dependent on the particular combination of fields and domains that they have interacted with, mediated through their social origins, the cultural scenes they are located within, their age, gender, and any other personality determinants that could have an influence on their artistic taste. Many DIY producers have hard drives full of half-developed material that has stalled, as for whatever reason the flow of creative excitement dried up, leaving the work suspended between production stages in a state of being partly written, partly arranged, and partly mixed. It will likely sit there waiting for a new stimulus to come along, such as an appropriate new commission for which the work can serve as a starting point, or a new inspirational moment generated somewhere else in their ongoing creative practice that motivates them to impose a structure to the work, commit to the core message, write the lyrics, and finish the damn thing.

The blurring of the staged production process is not new or specific to DAW-based production practice. Even before the widespread adoption of the DAW, hip-hop producers would create complete beats that contained specific hooks, rhythmic arrangements, timbral relationships, and mix balances long before a structure or lyrics were considered. A beat would then be circulated in hip-hop communities on CDs, DAT, or cassettes in the hope of inspiring an artist to create a song to it (Alexander 2007). Even in genres more associated with a 'song first' approach, such as guitar-based pop or country and folk, it had long been the practice of many songwriters in the days before the DAW to use four-track cassette recorders like the Tascam Porta 1, or later multi-tracking standalone workstations like the Yamaha AW16G, to fix arrangements and timbral relationships into specific ideas and meanings early in the compositional process. Zak cites Sting here, who wrote in 1983, "I make the best demo I can, singing and playing guitar, bass, piano and drum machines. I get a good version so that people will see its potential" (Zak 2001, p.27). It is fair to surmise that Sting wanted his collaborators to experience the song as he envisaged it, presenting it complete with the emotional content and meanings carried in the relationships between instrumental parts and tones fixed in the arrangement. His collaborators would hear the fully arranged song and attribute meaning to it in their own minds (Barthes 1977) that could be assessed and discussed so that decisions could be made to progress the song. In essence, the collaborators perform the duty of a test audience for the song, not just as lyrics, melody and chords, but as a fully written track. Well before the DAW it was a common anecdote amongst rock communities that specific sounds and performances from a multi-track demo could not be recreated or bettered in the studio production stage and so were lifted directly from the demo tape to survive into the final version. The demo part was deemed to have a particular inimitable quality that we could reasonably label character: something special that happens when a musician is performing the part for the first or nearly first time, before they become overly familiar with it and the performance loses its edge, straying close to recital when their fingers move across the instruments with practised ease rather than the concentrated, energised focus achieved when the part is new to them.

This aspect of studio performance has been commented on by Daniel Lanois during his interview with Massey (2009). He describes how Emmylou Harris has never bettered her original vocal take from the guide tracking stage on another day:

> That's where it all gets mysterious, because it's all beyond measurement; if it's the same mic and the same vocal chain and the same woman in the same room, then it should be the same ... but it's not. That's where you get into psychology and spontaneous combustion and freshness and all those aspects.
>
> (Lanois in Massey 2009, p.22)

As well as these difficult to quantify aspects, the practicalities of writing at home and recording in an expensive studio produce other characteristics that can make a demo part more desirable than a polished studio part. In a lot of projects, the demo parts are recorded with less effort put into the means of capture so as to get the ideas recorded quickly, such as using the first microphone that comes to hand, recording in a noisy environment, or with a substandard instrument. However, these can be the very things of happy accidents that can occasionally produce the most characterful, creative parts.

Perhaps a key difference with the contemporary DAW is the functionality to use the exact same DAW session used for writing and being able to take it into the studio for the production process as a blueprint for tracking the final, more collaboratively scrutinised performances and tones. This makes it easier for characterful demo parts to survive into the final version but could also constrain the creativity of the collaboration by presenting the easiest option of replacing the demo parts one by one in an overdubbing process that favours the songwriter's exact arrangement ideas, limiting the potential part-writing contribution of the other musicians. Such a production model also limits the reactive and dynamic performances achieved when musicians record their performances simultaneously and is in danger of producing sterile music that lacks a performance spark and fails to deliver the desired project meaning, aesthetic, or band persona. The producer should be aware of such dangers when designing the production process or creating the Production Habitus.

As the technology allows more and more compositional freedom (songwriting-wise and track writing-wise) to dip into any stage of the production process without the need to have completed previous stages, perhaps the value of conceptualising a linear staged view of the production process is as a self-reflective tool for the producer–artist. It doesn't represent an actual timeline of progression, but it does provide a structure to the record writing process that can help identify boundaries, such as an end point to an activity, and maintain a sense of positive development and progression towards a goal. Rather than representing a series of dependent tasks in a temporal process, the model can offer the producer a framework with which to evaluate their work and their production culture (or carefully constructed Production Habitus), focusing on areas of quality control and ensuring room for collaborative creativity to flourish. We can still use adaptations of the linear model as tools to zoom in to identify key production decisions and scrutinise the collaboration at each point, considering the flows of influence and action throughout the production. However, in placing a production as part of a wider system that includes interaction with an audience, an alternative model is required to better examine the development of artistic persona through a cyclical production process.

The Production Cycle of a Producer–Artist

If we set our interrogative focus at the level of a single project by a producer–artist, we can map out a cycle of record production that broadly describes the creative production process undertaken and how it results in a product with a distinct character (Figure 11.2). This would allow us to track the persona of the producer–artist through a project production cycle, examining where and how production personality is formed, whilst acknowledging that the transition between the stages described is fluid and complex. The persona will track back and forth between the stages, being affected by anticipated happenings in later stages and considerations from earlier stages, rather than step smoothly from one to the next, but the general progression will occur clockwise around the cycle.

The model of a Production Cycle illustrated in Figure 11.2 articulates a holarchy, or a hierarchy of holons (McIntyre 2019), and opens up the various stages of production for scrutiny of their constituent parts to consider where intended messages are formed and decided upon, how they are expressed through the three layers of record composition proposed by

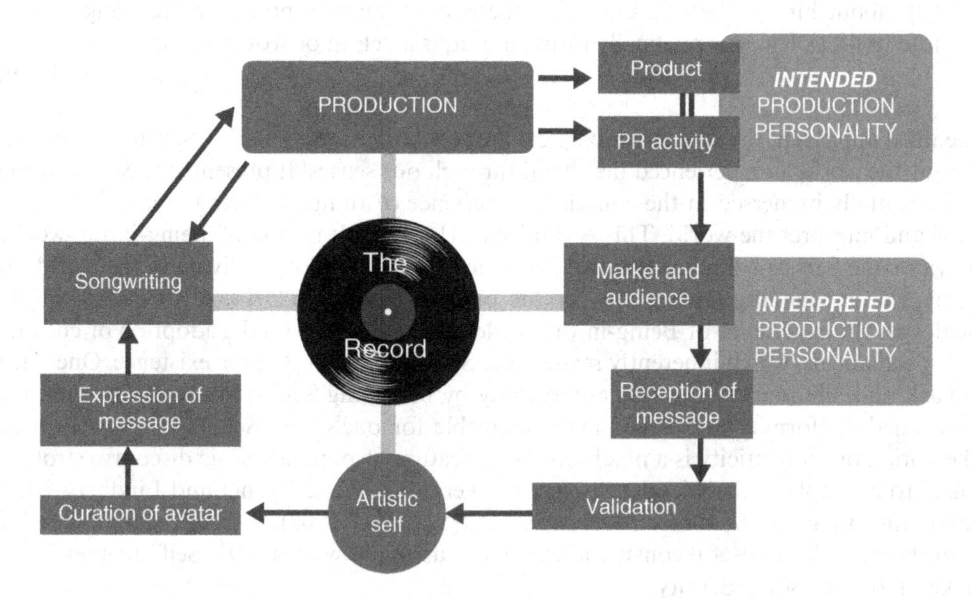

Figure 11.2 The production cycle of a producer–artist

Zak (2001), and how they stimulate meaning in the perception of the audience. It assumes that the artist has some back catalogue and defines key stages in the music production process where artistic identity is affected. The 'record' in the middle represents the artefact whose creation is the focus of the production. This artefact, whatever the format, will ultimately carry the artwork and its embedded messages to the audience. Any collaborators' personas will share this vehicle to get their artistic contributions out into the world and it is the producer's ultimate responsibility that this is achieved as successfully as possible.

We can start to examine the above Production Cycle from any point as it represents an ongoing process, including previously produced work and works to be produced in the future. I have chosen to start with the artistic Self, winding backwards through the cycle slightly to outline how this entity is partly formed from experiences of previous works, the market's response to it, and the subsequent levels of validation experienced by the artist. We will examine the identity of the artist, and how it can be understood as a multiplicity of Selves.

The Artistic Self and the Curation of Avatars

In his exploration of Heideggerian notions of self, Escudero (2014) identifies three main multidisciplinary contemporary approaches to understanding the Self. The Functionalist and Buddhist approaches discount the notion of the Self, suggesting instead that ongoing cognitive processes of self-representation are being mistaken for a distinct whole entity. The second approach, the Hermeneutic approach, views the Self as an evolving life-long project in which we construct a narrative Self through the stories we tell ourselves and others, and the stories others tell about us:

> In other words, our actions achieve intelligibility in a narrative sequence which provides insights on our character traits, the goals we pursue, the values we endorse, and our different ways of living. In the frame of this narrative and temporal sequence, the story of a life continues to be reconfigured by all the truthful or fictive stories that a subject

> tells about him- or herself. Our life undergoes a constant process of reconfiguration. Life itself, as Ricoeur poetically formulates it, is a "cloth of woven stories".
>
> (Escudero 2014, p.9)

The third approach, the Phenomenological approach, links the sense of Self to our experience of the world as experienced first hand through our senses. It presents the Self as being fundamentally immersed in the conscious experience of things, as being a part of how we sense and interpret the world. This is similar to Heidegger's notion of Being-in-the-world. He views the Self as being inseparable from the experiential flow of living in the world, of being "the unchanging inner reality of the person" (Escudero 2014, p.10). According to Heidegger the experience of Being-in-the-world inevitably includes the adoption of cultural and social norms, which inherently renders the Self into an inauthentic existence. One must embark on a life journey towards authenticity by becoming Self-aware, breaking with the entrenched conformity and becoming responsible for one's own Self (1967 pp.166–168). The notion of authenticity is a much-discussed feature of popular music discourse strongly linked to concepts of artistic identity (see Shuker 2017; Weisethaunet and Lindberg 2010; Barker and Taylor 2007; Butler 2003; Moore 2002; McGee 2007), so Heideggerian ideas of an authentic Self are useful considerations in discussing how an artistic Self emerges in the make-up of a person's identity.

In considering the concept of 'author function', Foucault recognises the existence of several Selves at play simultaneously and that our conceptions of the author operate on them as a plural entity, rather than on a distinct individual (2011 [1969]). In his discussion of signs within a text that refer to the author, such as first-person pronouns, he demonstrates that they refer not to the actual individual, "but, rather, to an alter ego whose distance from the author varies, often changing in the course of the work" (2011, p.533). Whilst Foucault was focusing on literary works, this aspect of author function is useful in discussing song production, where the first-person character in a song is commonly a fantasy extension of the songwriter or artist playing out a narrative or expressing distilled ideas in service of the song's message. Artist alter-egos are a mainstay of popular music and plenty of artists such as Prince, David Bowie, and Eminem, have used theatrical alter-egos to widen their creative expression, explore different areas of the human experience and keep their output fresh. However, in most produced songs the voice that is present in the work comprises several more subtly affected versions of the real individual singer–songwriter. The notion of author function on multiple Selves becomes more complex as we consider the subtle layering of identity within the artist–producer who presents themself as the creator of a produced song. One Self is the voice or character in the song, crystallised in time and emotion and limited to the messages in the song; another Self is the scribe behind the lyrics, deliberating over the precise language the character will use; one Self is the musical arranger; another the audio engineer; and another Self is brought to the work by the audience in their previous understanding of the artist–producer from past works, journalistic writings or social discussions around the artist and their field. For Foucault, these entities can be encountered simultaneously, and the notion of authorship is distributed among them.

The concept of multiple Selves inhabiting the same body is discussed in philosophical writing and clinical psychology (Braude 1995; Rovane 1998; Gallagher and Radden 2011; Ryle and Fawkes 2007). The *Journal of Clinical Psychology* devoted its entire January 2007 edition to the notion, in which Dimaggio and Stiles explain:

> Each person comprises multiple facets or voices, each possessing its own characteristics, expressing different emotions, and taking distinct perspectives on events and social interactions. As the various parts surface, the person acts and feels differently, processing

> information differently (Tooby & Cosmides, 1992), anticipating different reactions from others (Baldwin & Main, 2001), and displaying new action tendencies and constructions of meaning.
>
> (Dimaggio and Stiles 2007, p.119)

This alternating of multiple Selves depending on the situation is a process which Cognitive Analytic Theory terms 'contextual multiplicity' (Ryle and Fawkes 2007, p.165). Although much of the clinical psychology research focuses on cases where the inner multiplicity of a person is extreme to the point of disorder, perhaps we can borrow more generally from these approaches to the Self in understanding the emergence of an artistic Self. The individual whole Self, as a being defined through its experience of the world through bodily senses and interpretive cognitive processes, becomes more self-aware by accumulating experiences over time and reflecting on them. From this position of Being-in-the-world it strives toward a more authentic existence by rejecting cultural paradigms more and more, formulating its own way of doing things to become an individual expressing itself in the world creatively. Through experiencing interactions with others and adapting to various roles in reciprocal social relationships (Ryle and Fawkes 2007, p.165) the Self becomes multi-faceted in order to best approach differing situations. It accumulates narratives about itself that reflect its thoughts, deeds and encounters, and defines its inner dialogue through repetition. It continually curates these stories and expressions in its interactions with the world and others to present as a Self with a unique identity, propelling its life along a trajectory of defining and redefining itself. Each individual's conception of who they are is developed in these processes, which result in a complex contextual multiplicity of Selves that is simultaneously a single person, able to reason from differing viewpoints and conduct inner dialogues between Selves. For Gallagher and Radden, "the confusion and complexity found in the usual psyche are a valuable part of human nature, suited to the moral categories and social institutions we cherish" (2011, p.561).

An artist then, as a person operating in the world, can be considered to represent a complex embodiment of their social and geographical backgrounds, ideologies, education, influences, tribal affiliations, relationships, career to date, and other defining factors acquired through life. Of the multiple Selves that develop in this person, the artistic Self at times steps to the fore and creates work that contains messages expressing elements of itself to the world, such as attitudes, emotions, narratives, and views that articulate the individual's experience of the world through their art. It can be seen as distinct from, or at least a variation of, the complete personal Self in that it is a selective concentration of it, a distillation of qualities that represent the individual's artistic journey and that comments on things from the selective viewpoint of the artistic Self. Parts of the personal Self are emphasised, de-emphasised, fabricated, or ignored in the emergence of the artistic Self, which is brought to the fore when engaging in their art.

The artistic Self itself can also be perceived as a contextual multiplicity. The Me that strides confidently onto a stage at a gig is a different Me to the one that hunches over a piano, chewing a pencil and agonising over a lyric. And it is a different Me again that connects various bits of equipment together searching for a particular tone. These Selves are not simply task dependent. The Me that worked on Solo Album 2 is a different one to the Me that works on Solo Album 3, partly due to the different thematic content of the works, and also due to the passage of time between projects and the inevitable personal and artistic development that has occurred. This newer Me is also changed in a manner more directly relevant to my continuing artistic output by the very experience of producing the earlier album, as I have lived through and learnt from the production and distribution cycle. I know how I felt about how the work turned out, and how it was received by the audience and the critics in the field, and therefore I can reflect on creative decisions from an informed position. I can embrace or ignore comments from Others, including inner-Others; I can alter production practices to better elucidate, or veil, my artistic messages; I can direct my time and efforts more efficiently toward aesthetic goals; and I can better consider the position of the audience from an earlier point in the production based on their reaction last time, either by carefully preparing the material to be more accessible to a particular genre listenership, or by hardening my stubborn artistic resolve that deliberately disregards the market's response. Importantly, I will be conscious of my emotional and rational responses to the reception the previous album garnered from the market. Not just if they liked it, but if it connected with them emotionally, or if they understood it. There is a need for the artistic Self to get feedback for its efforts, and to know that the artistic expressions broadcast in its work were received by another. Such feedback can come from anywhere in the field: press reviews, personal messages, radio plays, social media likes and comments, inclusion on playlists, or interpersonal face-to-face conversations. By monitoring at least some of the impact of its work, the artistic Self looks for validation, and the level of validation and subsequent confidence as an artist is affected, for better or worse. Hence, with each new project the artistic Self is necessarily changed in an iterative cycle of development.

At the point of engaging in a new work the artistic Self will have an awareness of itself that is other to the version it is presenting to the world; the world-facing version is an avatar of the actual artistic Self, carefully curated to portray the artist in a particular way. We can cite Goffman here, whose influential book *The Presentation of Self in Everyday Life*, first published in 1959, used the metaphor of the theatre to suggest how an individual interfaces with the social world. Individuals can be seen as actors on the stage of social interaction, each performing multiple roles for the benefit of their audience (1990). He uses the term 'impression management' to describe how the individual presents an idealised image that is appropriate to the social interaction, exhibiting behaviours and a physical appearance that prevent the embarrassment of themselves or others (Crossman 2019). A psychological 'frontstage' and 'backstage' are distinguished with the frontstage being the performance area, where the deliberate version of Self is exhibited in the individual's interactions with the world; and the backstage being a private mental preparation space where the mask can come off and where mannerisms and artefacts of social performance are selected and practised, "It is here that illusions and impressions are openly constructed" (Goffman 1990, p.112).

Goffman's ideas have been used in more recent years to examine how individuals manage the wealth of data about themselves online and in social media interactions (Bullingham and Vasconcelos 2013; Crossman 2019; Hogan 2010; Jenkins 2010). The concept of avatar construction through a process of data curation is widely discussed, in which the individual presents an online self greatly enriched by photos, videos, biographical postings, selected web articles and memes. Bullingham and Vasconcelos' study (2013) found their participants' online personas could be embellished or exaggerated for the purposes of creating a more interesting social narrative. They might split the online Self between social media platforms to present several bespoke, slimmed down versions, each designed to best fit the particular platform. Or they might adapt their online persona to conform to the conventions of a particular online social space. Hogan (2010) distinguishes two types of self-presentation in the online world: 'performances', which refer to live online interactions with others; and 'artefacts', which involve the exhibition of curated digital objects that represent the Self to the online world. Goffman's front/backstage analogies can extend to this contemporary online presentation of Self, in that social media spaces are the frontstage; and the private user area of a social media portal, the desk area where the individual works on their personal computer, and the psychological space where language and actions are considered before enacting, are backstage.

There are obvious commercial reasons for an artist to project a particular image online, linked to their audience's expectations and any mythology they want to create around their artistic persona. We have seen how listeners construct an idea of a recording artist by combining the physical characteristics suggested by the sound of their voice on record with whatever other associated media they have encountered (Moore 2012). For the majority of artists, this other associated media will be encountered in online environments, dominated by social media. It follows, then, that Goffman's notion of 'impression management' is at play in the artist's frontstage and backstage actions in their management of their online presence. The artistic Self presents an online version of itself, or avatar, that projects to the world an idea of the artist that is desired. It does this in its performances in online interactions with others, and in its curation of digital content that populates its online spaces. The topics discussed, views expressed, and tone of voice evident in online conversations, combined with the photos, videos, design work, links, and memes exhibited in online spaces controlled by the artist, contain a wealth of meaning that is interpreted by the audience into an idea of who the artist is and what they are about.

This active curation of an online 'social' avatar is a continual activity in the lives of most DIY artists. However, in the early stages of a new project the artist is focussing on particular themes, concepts, and ideas that will define the new work. The artistic Self is exploring, considering the new subject of its artistic attention from various angles, searching for a voice with which to comment on it from. It is engaged in conversations with its inner-Others, defining a perspective for the new work, and allowing different inner qualities, stylings, attitudes, dispositions, and viewpoints to amplify or diminish until the most appropriate incarnation of the artistic Self is settled upon. It is preparing itself for the coming project. This process could be taking place without the artist deliberately controlling it, or even being aware of it happening. Unconscious responses to Being-in-the-world could be shaping the artistic Self in a particular way, building up over time until the artist feels a need for a creative outlet. It could also happen quite quickly following a commission or other external stimulus to start a project, but there is commonly this need for an incubation of the artistic Self; a reflective period in which the artist focuses on their artistic Self, favouring it over the many other Selves and reconnecting after a period of inactivity in this particular art, however short or long. An accelerated version of this activity can be found in actors as they enter the final minutes before stepping out onstage portraying somebody else. Photographer Simon Annand has focused on this moment in actors' work, photographing many actors over a 25-year period (Annand 2010; Thomas 2019). He shot his subjects during the 30 minutes before going on stage as they underwent the process of moving into another Self and literally stepping from a private backstage to a public frontstage, where they met the audience as someone else:

> Whatever theatre actors do during the day, each evening they go on stage to give a performance as 'somebody else'. The dressing room is a physical space that allows for concentration and privacy so the psychological negotiation between the actor and this fictional character can take place. When 'The Half' is called over the loudspeaker backstage, it is the start of a 35-minute countdown to facing the audience and there is no escape.
>
> (Annand 2010)

Although the 'character' they must adopt is 'fictional' and indeed written by the play's author and developed collaboratively with the director and company, the analogy with a recording artist negotiating with a version of their artistic Self is clear. They are preparing and operating a frontstage character, or avatar, from the psychological backstage of the inner Self. Many of Annand's photographs reveal intensity, gravitas, and concentration on the faces of the actors as they do their inner work. They also reveal the intimacy of the backstage space that they each have arranged for themselves to do the work in, which we can consider through the lens of the Production Habitus when discussing production environments (see Chapter 3).

The 'Curation of Avatar' node in the above music production cycle (Figure 11.2) represents this process of the artistic Self redefining itself for the coming project, forming an 'in-project' avatar. This is distinct from the audience-facing 'social' avatar as it describes the version of the artistic Self that is socially and creatively engaged in the project, who will explore the themes and express itself through every avenue of the artform. It is the version that the collaborators will encounter, the artist at work with their materials, more informal with their appearance and behaviours away from the views of the audience. In Goffman's dramaturgical approach we can say that the 'in-project' avatar performs on the frontstage in all social environments to do with the creation of the project, in the studio, in the rehearsal room, and in communications with collaborators. The 'social' avatar performs on a more polished frontstage for the wider online audience outside of the project's inner circle and is more tightly curated for its largely PR role. Both versions of the avatar share a backstage space, in which the artistic Self can prepare behaviours and social materials for them, unobserved by any audience. The avatars will develop over time as the artistic perspective for the new work becomes more established, but they receive a step change at this point in an artist's production cycle, when the themes, contexts and Production Habitus of a new work are being defined. The 'social' avatar will lag behind the 'in-project' avatar in its design, aesthetic and demeanour, not fully catching up until the project is in its later stages and the ideas, themes, overall sonic impression and artwork coalesce into a tangible product character that will be announced to the world in the promotional activity surrounding the release.

I have the notion that moments of intense human connection between artist and audience occur when the door separating the frontstage from the backstage is momentarily blown open, revealing all the mess, complexity, and truth of the human experience in a flash of deep interpersonal communication. Such moments are likely to endear an artist to their audience, and to encourage attributions of authenticity, but the artist–producer must be wary to keep them brief or risk being overly sharing and off-putting to the audience. Backstage only holds excitement and mystery so long as it is only ever glimpsed, and rarely visited.

12 'Silver Glass'

Re-production

Christopher Johnson

The previous chapter presented the producer–artist as a multiplicity of Selves that are curated into an artistic persona during the production process. Even when physically working alone, we have discussed how collaboration occurs with inner-Others as the artist engages with various interrelated systems and domains as they create their work. This chapter examines these ideas and the Cycle of Production through an auto-ethnographic case study. It analyses the Production Habitus and its impact on the final artefact, the development of an artistic Self through the Cycle of Production, and the effect of internalised Others as the artist makes their creative decisions whilst producing music.

Artistic Self in the Cycle of Production

I was in the middle of producing an album for the band Halo Blind in which I co-write and sing the majority of the material. There had been a break of a few months in the activities of this album project as the world first locked down for the Covid-19 pandemic in early 2020, and the disruption to the Production Habitus meant I lost momentum and creative focus whilst dealing with the other professional and personal matters the pandemic created. I needed to reconnect with the project and with my artistic Self in order to get back into a creative mindset, and to stimulate the part of that contextual multiplicity that considers the audience's position deeply when assembling tones and performances into messages that will be understood by others. Before committing to edit and performance aspects of the ongoing Halo Blind production, I felt the need to wind back through the production cycle, reacquaint myself with the continuing narrative and trajectory of thoughts that define and redefine my artistic Self, and seek artistic validation and authentication from an audience (Ryle and Fawkes 2007, p.165). Removing the barrier between front and backstage could be a way to achieve this.

My most well-known song is called 'Silver Glass', which I originally wrote and produced for the classic prog rock band Mostly Autumn on the album *Heart Full of Sky* (2007). It was the first song I sang lead vocals on for the band and has since become a mainstay of the band's live shows, and it has been sampled many times by hip-hop and dubstep artists around the world (for a few examples on YouTube see MrSuicideSheep 2013; The RealHipHop Channel 2010; SOURDIEZELDVD 2009). The song was originally written to comfort my mother when her sister was terminally ill in hospital, and it seeks to offer comfort and solidarity to those in similar situations. During the height of the Covid-19 pandemic it was prohibited to visit sick relatives in hospital, denying many families the chance to say goodbye to their loved ones. At the same time people were looking online for alternatives to attending live music events for their cultural interactions. It felt right to produce an emotionally raw, 'live' version of this song to reconnect with the audience and perhaps offer those families in our fan community who had sick relatives an element of

DOI: 10.4324/9781351111959-14

connection and solidarity. My intention was therefore to use the audience's familiarity with 'Silver Glass' and their existing understanding of its meaning and its status in the Mostly Autumn repertoire, and to site this within an authentic, stripped-down re-production of it to promote a sense of community, connection, and solidarity in our fanbase at a socially distanced time. In doing so I hoped to refocus my artistic Self for a period of further creativity by reflecting on my past works and stimulating my sensitivity to the position of the audience. I would record it live, film the performance, and offer the video to the Mostly Autumn audience via YouTube.

Creating the 'Silver Glass' Production Habitus

The logistical creation of the Production Habitus for this project was primarily personal and psychologically internalised, as I would be the only person physically involved in the production. This allowed me to go about the production being mindful of my own reactions and thoughts, able to make arrangements with the physical environment, project themes, and motivations that would optimise my physical and psychological performance state, bringing about the best conditions to perform under. The only physically present human collaborators in the production were my partner and daughter who contributed a vital element in creating the Production Habitus by agreeing to leave the house for an hour. Whilst a seemingly trivial contribution, they understood my need for privacy when performing emotional material. Even if they sat completely silently in another room of the house, my awareness of their presence and the possibility of them overhearing my vocal attempts would have affected the way I sang. Another key factor in creating a Production Habitus around this mini project was formulating a meaningful social and artistic intent, an artistically valid reason for doing it that would motivate creativity. We have explored in Chapter 3 the importance of project definitions, particularly when the project has a social aim, and we have seen through Amabile's Intrinsic Motivation Principle of Creativity (1983, 1996, 2013) that creativity is greater when participants are intrinsically motivated by the very act of performing the task at hand. Extrinsic motivation, such as the promise of reward or the prospect of being observed, is detrimental to creativity. At a time when a lot of musicians were streaming live performances from their homes quite casually, my less-than-confident performer Self needed the motivation it gets from feeling the work has deep meaning and social value. This is a trait absorbed from the Mostly Autumn doxa (or 'way of doing things') and the aura of deep meaning would help the work fit into the Mostly Autumn domain and be accepted by the audience. The sensitive nature of the social intent, plus the emotional content of the song, plus the fact that I had never played it before with solo piano accompaniment, created a fragile quality to the cognitive processes of the Production Habitus that I wanted to be present in the finished artefact. This kind of fragile performance can feel self-conscious and vulnerable as the artistic Self does the psychological work of getting into the appropriate performance character who can communicate the messages and the musical work of developing the vocal delivery, and so I'm grateful to my family for leaving me to it. I wanted 'fragile and vulnerable' to make their way into in the finished product but not 'self-conscious'.

We agreed a time in the afternoon when they would go for a walk for an hour, the maximum amount of time deemed appropriate by the government for outdoor exercise at the time (Clark 2020). The duration is an important detail, as it is an external factor that dictated the production process. I would need to get the technical preparations completed enough beforehand so that the hour could be spent focusing on getting into the right psychological performance space and then recording the actual performance. The time of day was another important factor as the performance time needed to consider mealtimes, energy levels, and vocal condition. I know that I can't sing very well for about two hours after a meal, and I perform best when I'm hungry. My vocals are always best in the evening after the days' worth of activity has warmed them up, but this wouldn't work with the household routine so the afternoon was the compromise. I would skip lunch and set up the studio before they left so that I could maximise the window I had to perform in.

The song was recorded live at my project studio desk and filmed inexpertly from three angles by the main desktop iMac, a phone propped up against one end of the keyboard, and a laptop positioned at the other end. The video cuts between these three angles, with the iMac front-on view filtered to black and white in the editing software. The performance position at my studio desk is the closest physical representation of my artistic Self's backstage area, and I am wearing whatever scruffy t-shirt I pulled on that day, rather than the more formal black shirts the band tends to wear for our quite lavish projection-heavy stage productions. I didn't tidy or light the room for the videos, but I did close the curtains to reduce some glare and give the performance a more intimate, closed-in aesthetic.

Backstage and Authenticity

By revealing this genuine 'backstage' to the audience I am showing them 'behind the scenes at Chris' studio' images that maybe one or two people might be mildly interested in, but more importantly I am briefly removing the barrier between physical and psychological frontstage and backstage to communicate more directly with the audience in a way that is perceived to be unmediated by the usual gloss and packaging of produced music. For them to be able to relate to this performance and for it to have meaning for them, they would need to believe in it as an authentic artistic expression and this factor had an effect on the production decisions. The audio set-up was a simple vocal mic through an LA-610 channel strip going into Logic X with a Lexicon reverb, and I'm playing a Native Instruments piano. To keep the aesthetic 'live' no mixing at all happened after the performance. I had a few rehearsals and then recorded a couple of takes and picked the best one. The finger clicks at the start of the video, originally done to allow me to sync the audio with the footage from the three cameras, were left in as a cue to prompt notions of a live unedited performance. The aesthetic intent of the high-quality audio but pretty ropey video was to produce an honest and authentic version of the song that would help communicate its message, slightly re-purposed for the Covid-19 lockdown situation. I could have spent a lot more time on the video, researched how to make it look better, sourced better cameras, lit the room properly, or even staged a more attractive scene, but such efforts would diminish the message, effectively keeping the 'frontstage/backstage' division more intact, and potentially undermining the fragility at the heart of the performance.

These production decisions can be viewed through Moore's (2002) work on authenticity, which highlights the importance of the relationship between an artist and the audience, with the artist rather than the work being ascribed as authentic by the audience when certain criteria are met. His 'first person authenticity':

> arises when an originator (composer, performer) succeeds in conveying the impression that his/her utterance is one of integrity, that it represents an attempt to communicate in an unmediated form with an audience.
>
> (2002, p.214)

He also cites the argument made by Taylor (1997) that an expression 'is perceived to be authentic if it can be traced to an initiatory instance' (2012, p.213). For the re-production of 'Silver Glass' the core messages of comfort and solidarity and the artistic motivation in itself would hopefully be understood as having integrity. The exposed aesthetic of the live performance production and the Production Habitus designed to achieve it are comparatively streamlined and unaltered by the mediation of post-production processes. The distance from the initial expression of the message to reception by the listener is small, and indeed demonstrates 'an attempt to communicate in an unmediated form with an audience' (ibid.). Moore's 'first person authenticity' can also be considered in the means of distribution. The video was posted to YouTube, which allows the inclusion of contextual information. It was posted with this description:

> A live, solo version of Silver Glass from lockdown. I wrote this song for my Mum at a time when she was having to comfort her terminally ill sister through her final days. We recorded it with Mostly Autumn for Heart Full of Sky in 2006. Revisiting it now from lockdown thinking of all the families unable to visit their loved ones in hospital.
>
> (Johnson 2020)

Including this message with its biographical notes is likely to increase the ascription of authenticity as it prompts the audience to consider 'that there is an Other present who *means* what he/she says' (Weisenthaunet and Lindberg 2010, p.477). The listener's experience of the performance is informed by the description they most likely read on accessing it. The integrity of the messages in the production and the relationship between performer and audience are highlighted, and so seep through the listener's understanding of the work as they encounter it. By considering the audience in this way, I am collaborating with an internalised version of them, learned through my immersion in the Mostly Autumn domain. Moore's 'second person authenticity' occurs when 'a performance succeeds in conveying the impression to a listener that that listener's experience of life is being validated, that the music is "telling it like it is" for them' (2002, p.220). This idea was threaded through the intention of creating a production that would help people feel connected. The whole world was in lockdown, everybody feeling the strangeness, isolation and fear of the situation and unable to leave their homes, and it was very much an experience shared by everybody likely to see the video. The very act of producing something transparently from within the limited production environment of the home was 'telling it like it is' for the audience and therefore likely to be deemed authentic. There is much more to investigate about how production expectations and quality control in all media were affected by the lockdown, from TV panel shows being produced with every panellist in their own homes, to the innovative meta comedy series *Staged* (Dir. Evans 2020) which took place entirely in a video conferencing environment, but it would be outside of our focus here.

In these ways, the design of the Production Habitus was key in facilitating the right kind of performance and the desired aesthetic of the production, as well as ensuring the audience would perceive the core messages as having integrity, thereby focusing on my relationship with the audience and stimulating my artistic Self out of hibernation.

Structures of Internal Collaboration

I began the actual production by spending some time playing through the song on the piano, making adjustments to arrange the music for solo piano and vocal as until then it had always been performed with the dramatic instrumentation of a large rock band. At the same time, the vocal mic was set up and I was making adjustments to its positioning, and to gain, EQ, and compression settings on the LA610 channel strip that was driving the vocal tone. I was also crafting the Native Instruments piano sound I was using, and trying out several reverb and delay options, effectively creating the tones I wanted simultaneously with arranging and rehearsing the piece so that the performance could be recorded live with no mixing required afterward. This process followed a scaled-down version of the cyclic DAW-based Track Writing Engine from the previous chapter (Figure 11.1). The entire piece was not arranged, then rehearsed, then performed, then mixed in a linear fashion, but rather each aspect was being developed in parallel, with one sonic event inspiring a tweak to another until the final production timbre emerged as one: mix, arrangement and performance gestures and all.

We can use Chapter 3's discussion of Csikszentmihalyi's systems model (2013), and McIntyre's notions of interconnected, scalable systems conceptualised in holonic structures (2019) to examine my decision-making here, how creativity is happening in the process, and how the artistic Self is being affected as a result. McIntyre outlines the domain, field and individual of the systems model and then clarifies that creativity happens in not just one system but in the interplay between many interconnected, scalable systems:

> From this perspective there is an engagement with a deeply multifactorial, non-linear, interconnected and scalable set of systems which extend vertically, up and down, system within system within system. These are also deeply and horizontally interconnected with related domains and fields,
>
> (McIntyre 2019, p.70)

The notion of multiple nested systems creating an intricate network of intersecting domains, fields and individuals allows us to consider how creativity is happening through a process of internal collaboration within the artistic Self (see Figure 12.1).

I suggest that for the 'Silver Glass' re-production involving one physically present contributor, there are three vertical levels of systems worthy of scrutiny, each consisting of several interconnected horizontal systems.

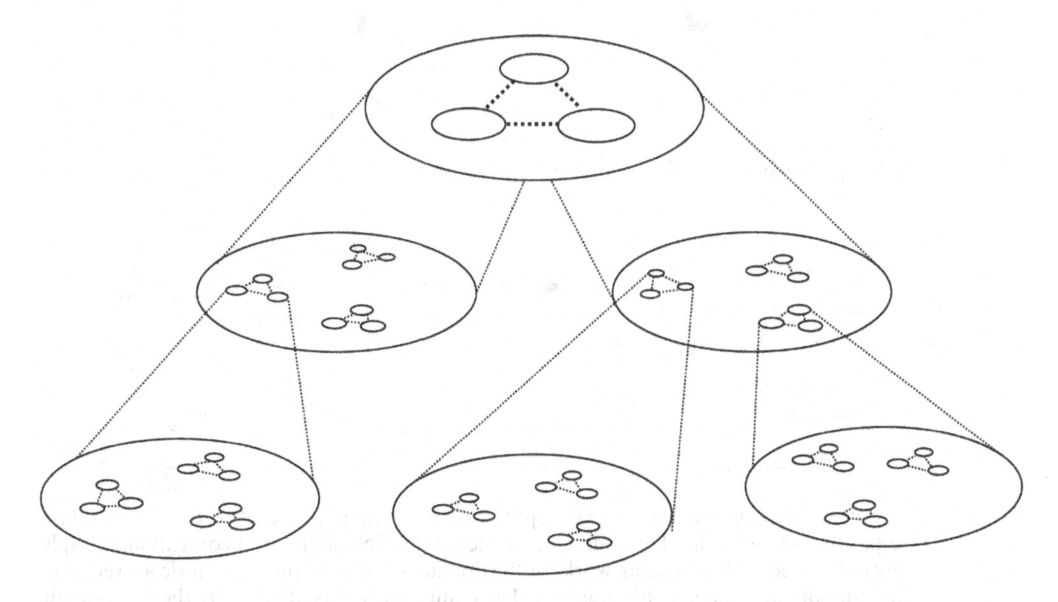

Figure 12.1 A holarchy of nested creative systems. One holon is one creative system comprising domain, individual and field, like the one seen at the top level in the diagram. Each system is in itself whole, whilst simultaneously making up part of other systems. This structure allows us to analyse the relationships between the systems, interlinked horizontally and vertically, as we drill down through the layers of the holarchy

For example, the rearrangement and rehearsal involved a lot of play and evaluation, with musical and performance ideas assessed against my own internalised domains. In his analysis of David Bowie's *Heroes*, Thompson (2019) demonstrates the systems model in action in the related domains of performance and studio engineering. He suggests that Bowie constructed his vocal performance by using his internal knowledge of the studio vocal performance domain (2019, p.161). That is, he evaluated his creative ideas against an internal notion of whether or not other singers and producers, as well as the audience, would accept them. Thompson goes on to examine how the producer, Tony Visconti, was simultaneously using his internalised domain of studio vocal recording to assess the more technical aspects of microphone selection, room acoustics and the creative use of equipment, in making his engineering decisions (ibid.). In the case of my 'Silver Glass' re-production, the process included a similar internal consultation with horizontally intersecting general popular music domains such as vocal performance, piano arrangement, and sound engineering, but with the individuals in the systems being different aspects of the same artistic Self, rather than different people. From the holonic perspective, this represents the top level of the structure (see Figure 12.2).

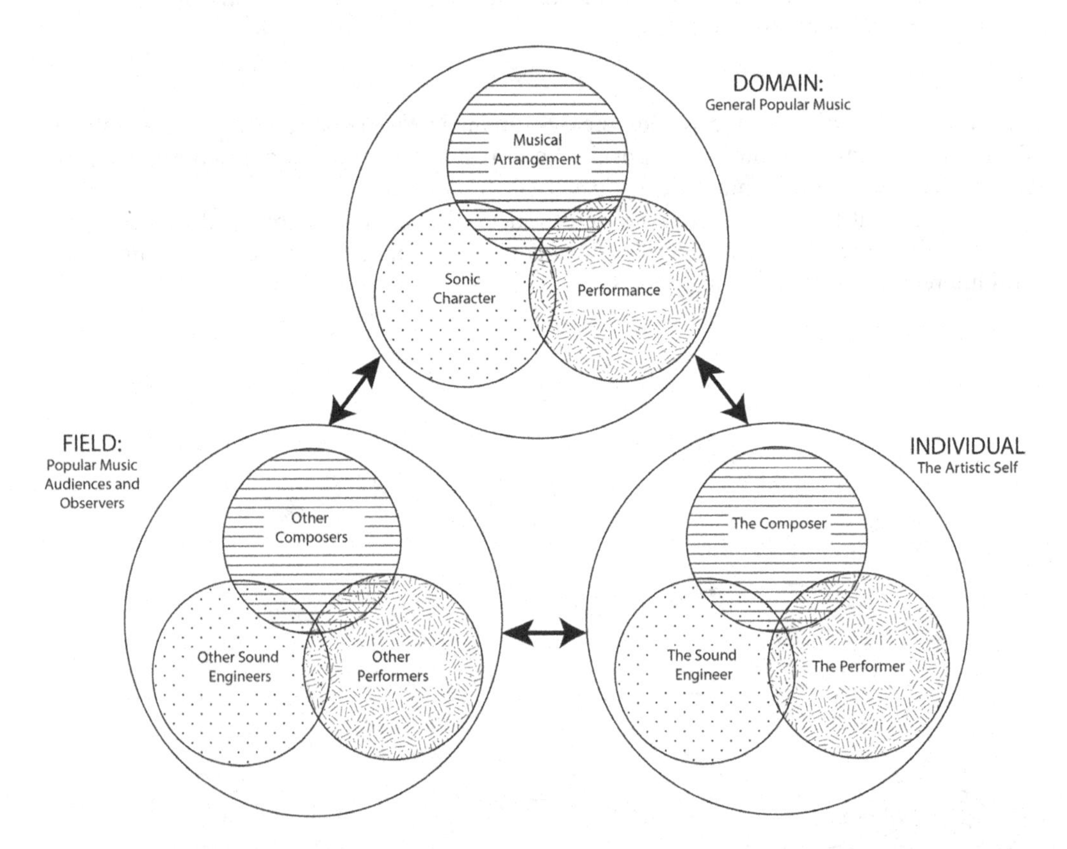

Figure 12.2 Interconnected systems of general popular music at the upper level of the 'Silver Glass' re-production holarchy. The Individual is the artistic Self, active as a contextual multiplicity concerned with preparing work for the domain of general popular music as well as its horizontally intersecting sub-domains. The composer Self is affected by the sub-domain of popular music arrangement; the performer Self is dealing with the sub-domain of pop music performance; and the sound engineer Self is concerned with the sub-domain of pop music sonic character. The field here is also multiple, as the artistic Self considers the field as a whole whilst simultaneously focusing on sub-groups within it

If we zoom in a bit to continue down the holarchy, it can be said that a number of micro-domains are nested within the large domains; other psychological processes at play that contributed to the evaluation and evolution of ideas. The first nested level below the more general system of 'all popular music' is the specific system of 'Mostly Autumn music', as I imagined and anticipated the approval or otherwise of the very niche audience who are familiar with the song already. For example, in an attempt to offer something refreshing to the audience who are the field in this nested subsystem, I decided to include the song's gentle coda, which is on the album but has never been part of the live band version as it diminishes the impact of the song's ending when played in a rock concert setting. I was confident the niche audience would recognise and welcome its inclusion due to my familiarity with the domain and my sense of the field. The "I" that was making these decisions, the individual in the system, was that version of my artistic Self that is active when engaged in Mostly Autumn activities. Having worked with the band for many years, this Self is well steeped in the domain of 'Mostly Autumn music'. It has worked on a lot of music, played many gigs, read many reviews, and in social settings where it needs to interact with fans and journalists, it has sent out its public-facing avatar – the curated Chris-from-Mostly-Autumn version of itself that it shares with strangers – and gathered audience reactions through direct conversations. This understanding of the micro-domain of 'all Mostly Autumn music' and the knowledge of its related field facilitates creativity at this niche level of the holarchy (see Figure 12.3).

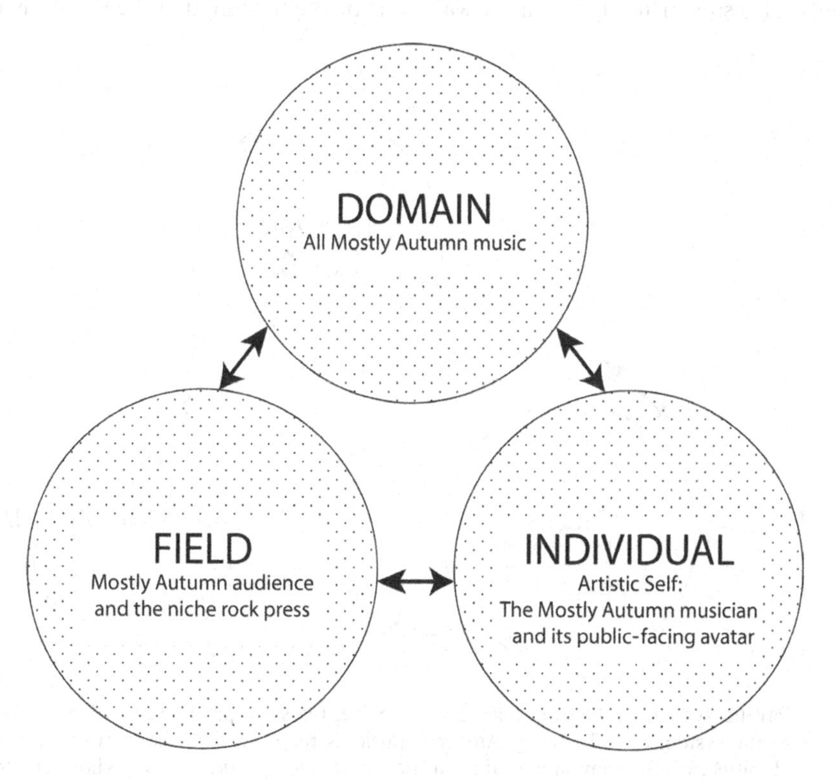

Figure 12.3 The niche system of 'all Mostly Autumn music' at the second level of the 'Silver Glass' re-production holarchy. The domain is defined by previously existing Mostly Autumn work, with the field being Mostly Autumn fans and the journalists that comment on the progressive rock niche that the band occupy. The Individual therefore, is the part of my artistic Self that is active when working with the band, and it public facing avatar

Parallel to this is the system of 'Mostly Autumn Production Habitus', which contains detailed knowledge of the social rules, technical habits, and production processes normally followed in the band's production activities (Figure 12.4). In this system, I am the individual band member creating something outside of the usual Mostly Autumn modus operandi, which involves the bandleader Bryan Josh's leadership, rough demo productions, quick rehearsals, and a specific tracking model in a specific studio. Here, I am working outside of this model to present something new to the rest of the band and associated production personnel, who are the field in this system, in the hope that they will happily be associated with the new production and it will be accepted into the band's collected works. Simultaneously, the hope is that my production process is accepted, which would widen the possibilities for the band's future output (see Figure 12.4). The existence of this internalised 'Mostly Autumn Production Habitus' system affected my musical and technical creative decisions. Much of the band's output is informed by a 'from the heart' aesthetic, which elicits ideas of the authentic expression of raw emotional themes, and this is partly achieved by a quick tracking process where takes are assessed far more on their emotional content than their musical or technical perfection. In the established 'Mostly Autumn Production Habitus' this is largely a function of the studio the band uses. Fairview Studio is built around an Otari Radar II recording system. Designed as a direct replacement for tape machines, the Radar system does not have the advanced editing capabilities of a modern DAW. On its release in the 1990s its converters were highly regarded by engineers moving from tape to digital systems as they achieve the high-quality warm sound associated with analogue systems. The workflow is also similar to tape in many ways, one of them being that the unit only offers a

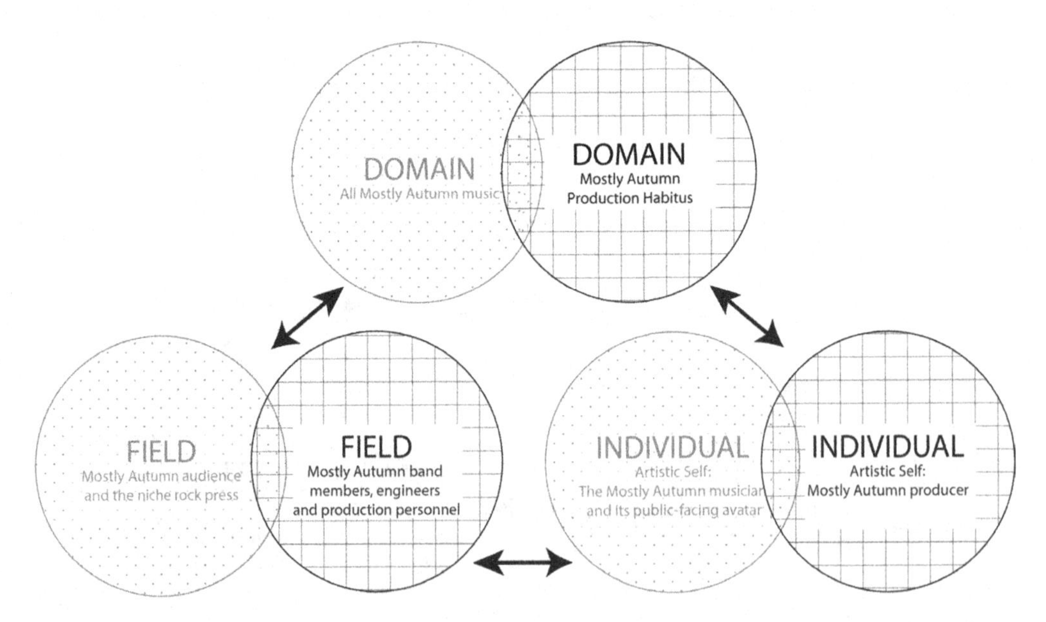

Figure 12.4 Parallel systems at the second level of the 'Silver Glass' re-production holarchy. Horizontal to the system of 'all Mostly Autumn music' is the system 'Mostly Autumn Production Habitus', whose domain contains all the social rules, production workflow, environments and ways of doing things that normally govern a production. The field is made up of band members, engineers, and crew that will assess my production methods, and the individual is the producer part of my artistic Self that normally works as part of the Mostly Autumn production team

single layer of undo (Robjohns 1999). This imposes the limitation of destructive recording whereby early takes cannot be kept, recalled, and compiled later as we would by using a DAW. As with tape, if you record over a take it's gone forever. In this way the limitations of the technology used in a normal Mostly Autumn production forces discipline in performance and decision-making in the studio. Once a performance feels right it stays, with little or no overdubbing or editing, and this classic rock production process imprints on the band's studio output. At home, I was using Logic X for my production, but my internalised domain of 'Mostly Autumn Production Habitus' and in particular the Fairview studio workflow directed me to limit my editing options and focus my performance. In the end I rehearsed until I was confident enough, then hit record on the Logic X and three cameras. I reviewed the take, noted a few areas to improve and performed the whole song a second time. This second take was better so I used it, with no corrections and no audio editing beyond cutting out the shuffling sounds before and after the performance as I go round the cameras turning them on and off. In this way I am making production decisions based on the usual Mostly Autumn doxa in order to imprint that particular essence from the process onto the produced artefact and have it accepted by the other Mostly Autumn personnel familiar with the process. Much of the rest of the Production Habitus, such as presenting a home performance, including just solo vocal and piano, an attempt at showing 'backstage', and the distribution method, are new to the Mostly Autumn way of doing things. They represent my novel contribution that I judge to be well-considered enough to incrementally advance the domain by opening up new approaches to how the band produces and presents material to the audience (Sternberg 1999).

Descending further into the holarchy, a deeper nested level involves a more personal consultation with inner-Others; mental traces of specific significant Others in the band. For example, although I played the piano on the record 13 years ago, the piano part has since almost exclusively been performed by the band's keyboard player, Iain Jennings, and it is his subtlety adapted performances of it that I am most accustomed to singing to in the frequent live performances. This influenced some of the nuances of the arrangement and performance as I felt it would affect both Iain's and the audience's acceptance of the production. Similarly, the band's leader, Bryan Josh, and second keyboard player Angela Gordon, tend to encourage me to sing with more vulnerability than I do in other scenarios, and their influence was at work as I practised the vocal, gaining confidence from the feedback I was getting from these physically absent and psychologically internalised collaborators to use a more fragile delivery than I do in the live performances. These influencers were not the actual Iain, Bryan, and Angela, but projections of them summoned into my consciousness during the creative process. They represented specific aspects of the real people built from my relationships, knowledge, and experiences of them through years of friendship and musical collaboration, but only existing in this particular form during my creative process producing a piano and vocal version of 'Silver Glass'. If I was working out guitar parts for one of his songs, another projection of Bryan would be involved, but it would be a modified one with a different set of criteria to assess ideas against. The systems model loses some significance at this level. If a domain, field, and individual are all psychological constructs within the same person, it is perhaps a less useful model in understanding creativity. The field is not in a position to admit contributions to the domain and so it cannot evolve. However, the projections of these significant Others borrow social and cultural capital from their real-world counterparts, as I found my inner dialogue responding in kind to their imagined reactions, and therefore they do have significance in the Production Habitus and their influence is audible in the finished production. The artistic Self is also affected by the influence of these unseen collaborators and can evolve at this level of the creative holarchy (Figure 12.5).

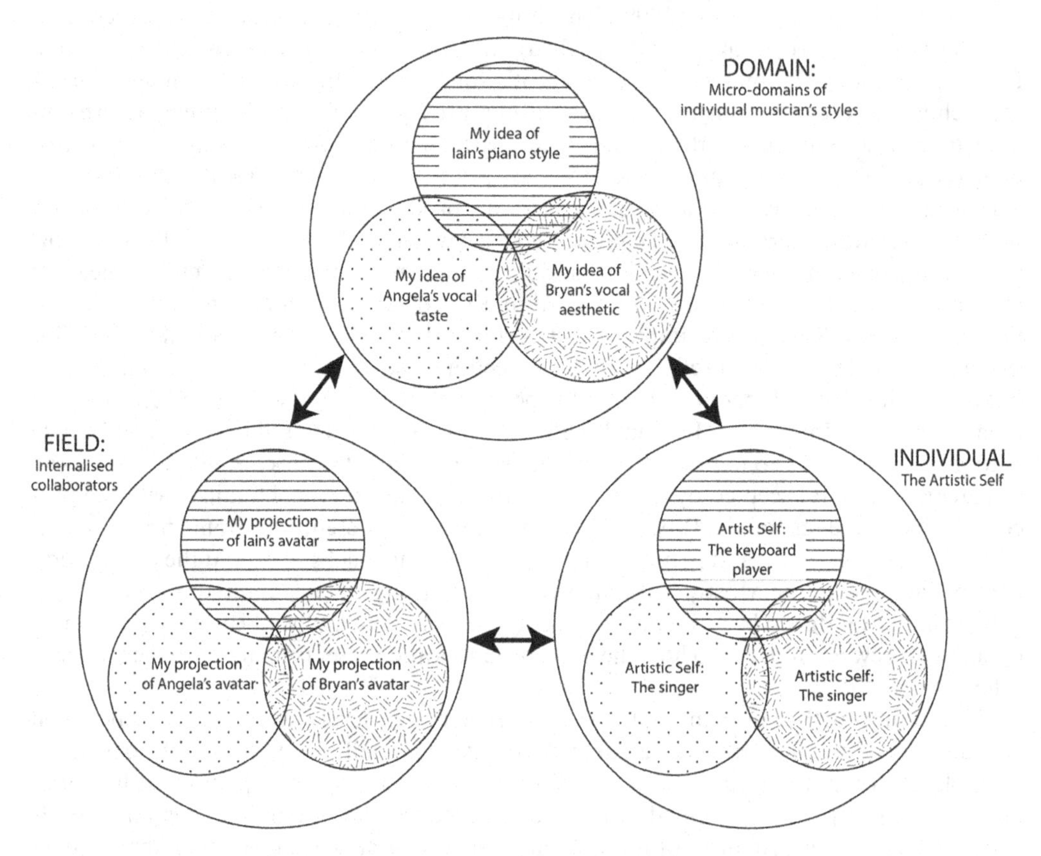

Figure 12.5 Interconnected systems of internalised collaboration at the third level of the 'Silver Glass' production holarchy. The artistic Self, both as singer and keyboard player, is affected by its previous experience of fellow collaborators' styles and dispositions, and its consultation with mental projections of them. The produced artefact and the artistic Self are changed in the process

Conclusions from 'Silver Glass' Re-production

In such ways the performance, production, and Production Habitus are very much the result of a collaboration, even when working alone in the physically and socially isolated environments involved in producing music in the Covid-19 lockdown. We could track the artistic Self through the Cycle of Production as it sought to reconnect with an audience and prepare for new creative work by manipulating the frontstage/backstage dynamic, both literally and psychologically. The Production Habitus remained a crucial consideration, even on a one-person project, as the project goals, physical environments, time of day, artist motivations, and social micro-culture all had an audible effect on the product. The social and cultural capital of the artist and the source material within the field of Mostly Autumn were consciously handled in a way to better connect with the audience, and therefore make the project more successful in achieving its aims. The perspective of the holarchy of interconnected systems models is useful to scrutinise the effects of internalised collaborators, although it gets messy as we delve deeper in the psyche of the artist. The concept of the Production Habitus is more useful at this level, as we can better observe the effect of cultural and social capital, and friendship, on the artist at work.

A final production decision to move this small project to its aesthetic and social goals was the method of distribution. It was uploaded to YouTube (Johnson 2020) without fanfare and the link posted on the Mostly Autumn Facebook page. This group doesn't have many members, but the video had been viewed 477 times in the first 24 hours, and the positive comments left on YouTube and in the Facebook group suggested that it had achieved its aims from the audience's perspective. From my perspective the project successfully re-energised my artistic Self and helped me focus on aspects of performance, presentation, and production that consider the audience more than I had been doing working in isolation.

13 Play One We Know!

A Pub Singer's Struggle to Retain His Integrity Whilst Remaining Entertaining

Christopher Johnson

Introduction

As a songwriter who regularly entertains in a pub environment, I am frequently plagued by the sense of having sold out; a notion that by singing other people's songs I am denying my own creativity, contributing to the dumbing down of popular culture, distancing myself from musical peers and betraying my artistic conscience to take the easy money and keep the audience clapping and the gigs rolling in. Why should I feel like this when the job I am being paid for is to entertain, and the audience will always favour a song they know over one they don't? This sense that I might be performing in an inauthentic manner is the driving force of this chapter, and what requires examination is my own criteria for constructing authenticity.

The idea of authenticity is a central feature in popular music discourse, prized by musicians, debated by fans and analysed by scholars. Roy Shuker (2017, p.24) writes that authenticity assumes "an element of originality or creativity present, along with connotations of seriousness, sincerity and uniqueness," and Weisethaunet and Lindberg (2010, p.465) point to terms like "integrity", "honesty", "sincerity", "credibility", "genuineness", and "truthfulness" when introducing their ideas about authenticity. Barker and Taylor define the term 'primarily in opposition to "faking it"' and note "the notion that honest, raw, pure self-expression is the thing that matters" (2007, p.ix). Such concepts of originality, creativity, and self-expression are central to discussions about authenticity and form much of the common person's understanding of the term. Butler writes that "authenticity is constructed discursively: within musical communities, fans, critics and performers argue about what constitutes authenticity and why" (2003, p.1), therefore my audience's opinions, or at least my imaginings of them, must affect my understanding of my own authenticity.

As a modern freelance musician with a portfolio style career I find that my differing roles afford differing bandwidths for creativity and personal expression, which impact upon my feelings of authenticity. If we use Sternberg's Propulsion Model of Creativity (2006, p.96) we can map my musical occupations against his creativity types as a crude measure of paradigm shifting and how original each is allowed to be. At one end of the spectrum of my activities, as a songwriter and music producer, the possibilities for expressing an original voice are complex and limitless. I can choose to produce original challenging music, rejecting the current trends of whatever genre I am working in, or conversely write songs that reinforce genre expectation and fall into Sternberg's 'Replication' type of creativity. This is rarely a consideration whilst writing as the creativity type and genre framework of each song can only really be assessed retrospectively, but the point is that they are written in a mental space of relative artistic freedom where self-expression is valued over all other factors, and therefore I feel authentic. As a hired gun playing session guitar or keyboards in another songwriter's band, or as a composer responding to a client's commission, the creative possibilities change as effective collaboration within a stylistic framework becomes more important than raw self-expression, but although less frequent, paradigm shifting

DOI: 10.4324/9781351111959-15

contributions are still possible and still valued. I am contributing to original work through a creative process that allows for all Sternberg's creativity types to emerge and therefore I feel authentic.

At the other end of the spectrum of my activities, as a solo performer in a pub environment playing acoustic guitar and singing through a small PA system, my role is to entertain the people in the room. An assumption exists that the songs I will play should be familiar to a generic pub-going crowd, which limits the creativity types to contributions that accept current paradigms (Sternberg 2006, p.96). I want to play songs that resonate with me, regardless of whether or not they are in the public awareness, and therefore such an approach feels less creative and inauthentic. The importance of this feeling of inauthenticity in a performer cannot be understated. In their book *Faking It*, Barker and Taylor describe authenticity as an absolute, "a goal that can never be fully attained, a quest" (2007, p.27). They examine the effect the desire to be authentic has had on performers and whilst it can produce wonderful heartfelt music, such inner conflicts can be damaging and led to the undoing of Kurt Cobain, Richey Edwards, and Tupac Shakur in the 1990s (Powers 2009). "The question of authenticity in popular music is [...] fundamental to thinking about, listening to, and *performing* it as well" (Barker and Taylor 2007, p.xii) (my emphasis).

This presents our problem: is it possible for a performing songwriter to retain their sense of authenticity whilst still fulfilling their role as a pub entertainer? Is it artistically shallow for them to play it safe and limit their set selections to well-known cover songs whilst rarely performing their own original material, when their natural creative impulse is to work at propelling their art forward and expose audiences to material they haven't heard before? What would happen if they indulged their artistic reluctance to give the audience what it wants, or what they think it wants, and removed their imposed limitations to reject current paradigms and perform obscure, challenging, and original material?

Covering Our Tracks

To begin to answer these questions we must look at the songwriter's motivations in performing other artists songs in the first place. In the context of this study the financial benefit of sustaining an income through repeated bookings by satisfied customers is very important and carries its own integrity, but there are many more factors at play. There is a liberating appeal in performing cover versions. Kelly (2012) describes it as "a great opportunity to exercise the kind of artistic expression and wild abandon in performance that's not so easy when applied to one's own work."

In the 1960s bands such as the Rolling Stones and The Beatles validated their own authenticity by paying homage to artists they admired by imitating them and re-recording their songs (Weinstein 1997, p.141). This practice is dated in popular culture but still persists at the grass roots level where performers indulge the need to emulate their heroes by singing their songs; a practice also useful in trialling and developing performance styles in the amateur musician. Weinstein goes on to document how the occurrence of cover versions declined in the late 1960s:

> The modern romantic notion of authenticity – creating out of one's own resources – became dominant over the idea that authenticity constituted a relationship, through creative repetition, to an authentic source.
>
> (Weinstein 1997, p.142)

It is observed that such a historical progression is mirrored in the attitude of the developing singer songwriter whose focus shifts from copying their hero's songs to writing their own, and that songs learned during the early phases of their career can hold nostalgic charm for them. Performing them becomes a way of demonstrating their roots; something that seems in keeping with a folkloric notion of authenticity and something Weinstein includes as a postmodernist reason for performing covers (1997 p.146). She goes on to describe 'authentic appropriation' and what I feel as a performer is the key motivation for performing covers: "finding the original song to express the cover artists' views or feelings as well as if not better than anything they could write". This ties in with musicologist Allan Moore's 'third person authenticity', which we will look at in more detail later (Moore 2002, p.214).

In Weinstein's observations, the postmodern moment in rock music presents the past and its catalogue of songs as available for appropriation by current artists:

> If God is dead and all things are allowed, the God that was knocked off by punk was the myth of the individual, along with its master name: Authenticity. Musicians can now plunder the past with abandon.
>
> (Weinstein 1997, p.145)

Perhaps she was referring to Barthes' observations as he kills off the god-like Author and forefronts the relationship between the reader and the text (1967). In our situation we can see the Author as the original artist and Barthes' 'scripter' as the pub singer, disseminating the text like a shaman in an ethnographic society "whose 'performance' – the mastery of the narrative code – may possibly be admired but never his 'genius'" (Barthes 1977 [1967]). For Barthes, the modern writer can only create something 'new' by recombining what has already been written. These observations should comfort our anxious pub singer who can take note that even his most original songs constitute a collage of works written by those that went before, each element in his writing existing like a miniature cover version, and therefore his concern about performing cover songs can be viewed as merely a matter of scale. His authenticity then, intact regardless of the authorship of the songs he performs, is constructed in the mediation between author and audience.

The Audience: Who Are They?

We have seen that the relationship between performer and audience is central to the concept of authenticity. There are academic writers for whom authenticity is a quality *ascribed to* works rather than being inherent in them (Moore 2002, and Fornäs in Weisethaunet and Lindberg 2010, p.455). To feel authentic then, two conditions must exist: our pub singer's performance must be ascribed authenticity by the audience, and our pub singer in turn must ascribe authenticity to that audience. He must feel that they are competent in making their ascription. This presents the parallel questions of how does the audience measure the performer's authenticity, how does the performer assess the audience's competence, and how does the performer *think* the audience is measuring his authenticity? To examine these questions, we must look at the types of audience encountered.

Typology of Observer

There are three major groups that affect the performer:

- Familiar audiences: people who have seen them perform several times before;
- Unfamiliar audiences: people who have never seen them perform;
- Significant Others: a fellow musician; the person paying for the performance, usually a pub landlord or manager; partners; or potential employers.

Other factors are also considered, such as their behaviour, ages, subcultural tendencies, the venue, what day it is, and what time of the day it is. These factors shift and morph in the mind of the performer as they constantly shape their song choices and performance approach. For example:

- Weekend or midweek audience;
- Listening formally and attentively;
- Half listening but chatting also;
- Rowdy drinkers who are into the music;
- Rowdy drinkers who ignore the music;
- Quiet diners;
- Keen music fans or casual music listeners;
- Wine bar crowd or real ale pub goers;
- Rough age brackets: (18–30), (30–50), (50+).

How Does the Pub Singer Assess the Competence of the Audience?

In examining audiences, Alyson McLamore (2012) notes that, "listeners evaluate a performance on the basis of their blended personal and collective attitudes" but also that, "it is impossible for a performer to anticipate and respond to all the individual standards by which his performance might be judged." Nevertheless, the pub singer will continually try to 'read' the room and monitor the attendees to inform his song choices by responding to an assumed majority preference, based on how good he thinks they are at listening. For example, an attentive midweek audience assessed as being reasonably competent listeners will get more obscure and lyric-led material, whereas a rowdy weekend crowd, judged to be incompetent listeners, will get high-tempo recognisable songs. The interesting fly in the ointment is when a significant other walks in and the pub singer alters his selections to impress that one person. If it is the pub landlord a mainstream song will be chosen to snag the attention of the majority and demonstrate the performer's ability to keep the mood up and the beer selling; if it's another musician, judged to be a highly competent listener, an original or obscure song will feature to prove that even in the pub entertainment environment the performer is still a creative and artistic individual.

Using the typology of 'observer' we can approximate the processes taking place in the performer's mind. For example, they might think:

> It's mid-week, early evening. I don't recognise anybody here. It's a wine bar, most people are in gentle conversations and not really listening to me, a large group of suited men 40 to 60 years old are over there – some of them must be music fans, so I'll play some melodic mid-tempo not-too-obvious Neil Young song that they might just recognise but probably haven't heard played in a bar before.

The 3D graph below (Figure 13.1) visually illustrates some of what is happening in the performer's mind. If we plot energy in the room (y-axis) against familiarity of audience (x-axis) and how much attention there is on the performer at any moment (z-axis), we can see how they might make their song and performance style decisions. The darker spheres represent example scenarios created by typical combinations of these factors, and the thought bubbles show how the pub singer might arrive at their song decisions. Stefani suggests that different types of competencies exist relating to how successfully listeners can decode messages communicated in music (in Middleton 1990, p.175). 'High competence' is juxtaposed with 'popular competence' and the scope of variability between them is noted.

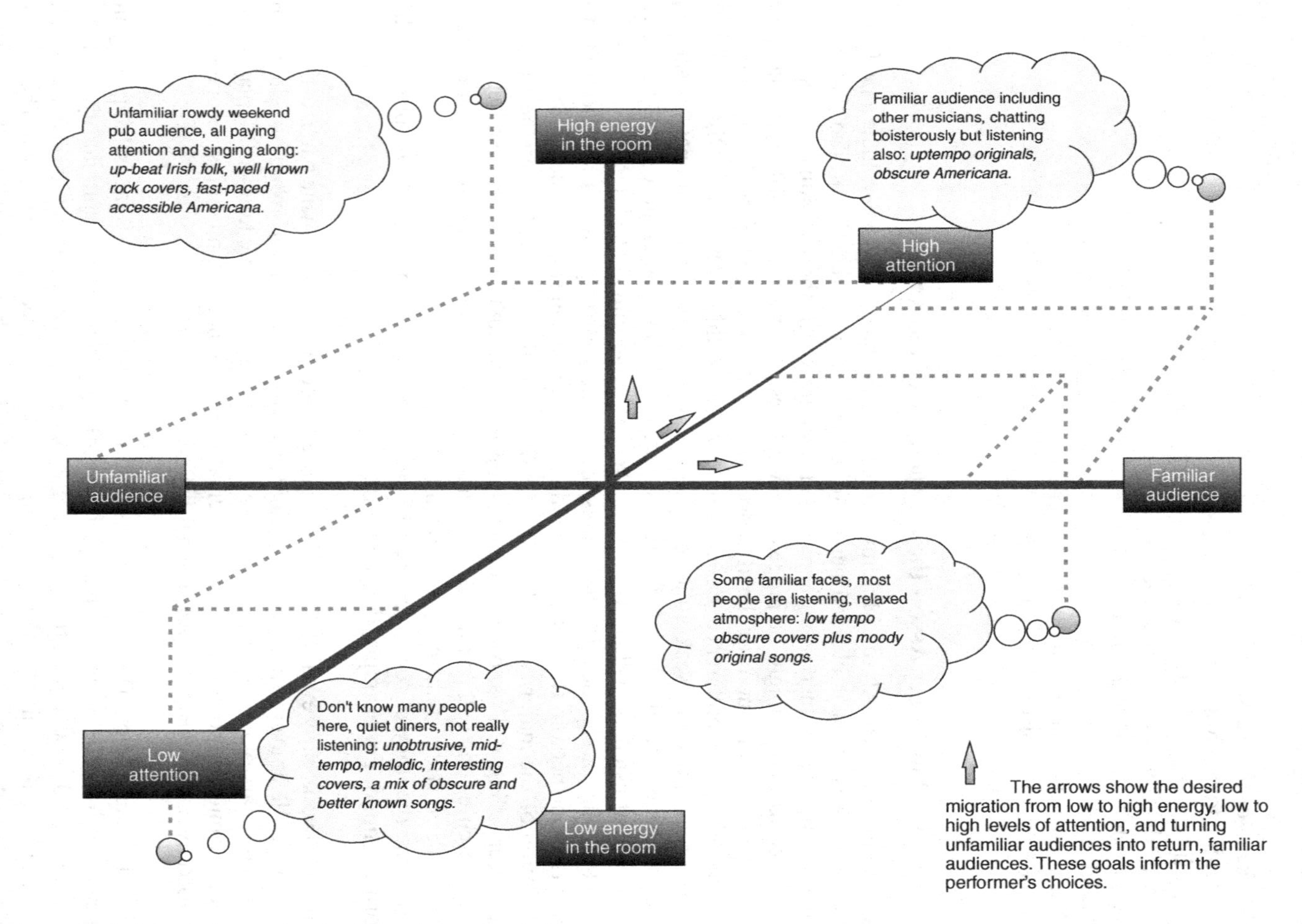

Figure 13.1 What is happening in the performer's mind

The pub singer judges his audience's ability to appreciate good art and then places them on a mental scale of competence.

From reading the assessments made by our pub singer it seems that in his judgement, audience members earn their ascribed competency by becoming more familiar, favouring a calm listening mode over a boisterous one, and flattering the artist with their undivided attention. They are rewarded with obscure and original material. In this model, the performer has become a monarch, only awarding his highest art to his most favoured subjects. Perhaps we can point to the pub singer's enculturation in the rock music world for this apparent arrogance and egotistical idealism, and his life-long exposure to the previously discussed mythologies in rock journalism. Only respected musicians are awarded such high competence that the pub singer will completely switch tactics to chase their ascription of authenticity and avoid the shame felt when playing unchallenging material.

Catering to the Audience

Can the performer be authentic in such a playing-to-the-crowd situation? According to Heidegger (1967, pp.166–168) one begins life inauthentically by being born into cultural norms and can only reach toward an authentic existence by realising this, becoming responsible for one's own self and breaking conformity. By allowing himself to be influenced by Others in the room our pub singer is, to Heidegger, behaving inauthentically. A paradox exists though, where the Other that the pub singer is reacting to is not the actual group of people in the room but a construct in his own mind. It is a projection of a multiplicity that represents his understanding of the audience. He empathises with the audience as a whole and makes assumptions on which to base his actions. In this situation where the Other is constructed by the Self, 'the Other would be a duplicate of the Self' and as such understanding is obstructed by all the filters of experience and 'possibilities of Being' that exist within the performer (Heidegger [1967], pp.162–163). This means any guess at what the audience wants results in the paradox: that in attempting to take the inauthentic position of making decisions based on the influence of Others in the room, the pub singer is actually becoming more authentic by only reacting to his own Self.

We need to address the binary of what the performer wants versus what the audience wants. Romantic mythologies surrounding authenticity cultivated in the journalistic field (Weisethaunet and Lindberg 2010, p.467) promote a view of fiercely independent artists 'keeping it real' and staying true to their artistic vision regardless of audience expectation. In *Faking It* Barker and Taylor (2007) demythologise such famously 'authentic' musicians and music by studying where such music originated. They find that catering to the audience has always taken precedence over personal expression (McGee 2007). There is a danger here: performers trying to play to the common popular musical tastes are accused of contributing to the dumbing down of the art, the 'cultural clogging' levelled at 'shiny floored TV talent shows' like the X Factor (Bragg in Plunkett 2012). In the Marxist view posited by Adorno, the performer would be expanding mass media and the 'manipulation of the social experience by the ruling class' by standardising his output and allowing the spiritual element in his music to wither (Llewellyn 2012, and Witkin 2003, p.11), hardly a position we can associate with the concept of authenticity. There is also an arrogance in making assumptions about what the audience wants. This needs to be avoided; the audience needs to be respected and ascribed an authenticity of its own by the pub singer if he in turn is to feel authentic in his art. Whilst maintaining an awareness of the energy in the room in terms of tempo and intensity, the best way for him to respect the audience might be to completely ignore his assessment of what they want.

How Is the Audience Measuring the Performer's Authenticity?

Mapping the Pub Singer's Situation against Allan Moore's Tripartite Theory of Authenticity (2002)

1. First Person Authenticity

Allan Moore asks *who* rather than *what* is being is being authenticated: authenticity is ascribed to the performer by the perceivers (2002, p.220). In his tripartite typology to sort how authenticity is ascribed, the 'first person authenticity' "arises when an originator (composer, performer) succeeds in conveying the impression that his/her utterance is one of integrity, that it represents an attempt to communicate in an unmediated form with an audience" (2002, p.214). In the case of our pub singer it seems that this type of authenticity can only be ascribed if they are singing a self-penned song and the audience are aware of it. The ascription may be stronger if the song has autobiographical content, and stronger still if the audience have prior knowledge of the singer's life, which is perhaps difficult in the pub environment where most people in the audience are unfamiliar. On a simple level, perhaps it can be achieved by announcing, "This is a song I wrote about being a pub singer."

Moore's first-person authenticity again highlights the importance of the relationship between performer and listener. Acknowledging the work of Moore (2002), Weisenthaunet and Lindberg set out another typology of authenticity (2010). They draw on Habermas' Theory of Communicative Action when they define their notion of 'Experiencing Authenticity' (Weisenthaunet and Lindberg 2010, p.477). This suggests that for a listening experience to be perceived as authentic "the recipient decides that there is an Other present who *means* what he/she says" (2010, p.477). These theories suggest that our pub singer can gain authenticity by addressing an audience between songs and building a relationship with them throughout the performance. The problem here is that this is a non-musical skill and the self-conscious performer can seem unnatural when addressing the audience without the veil of music. Happily, Weisethaunet and Lindberg allow us to be vague in our song introductions: "this Other may use indirectness as an aesthetic device, but, as long as the listener manages to reconstruct its intentions as sincere, any kind of performance can be found 'authentic'" (2010, p.477).

The Importance of the Spoken Introduction in Musical Performance

Certain performers can take this inter-song relationship building into an art form of its own. We have seen how it can help the construction of authenticity by revealing autobiographical details to a new audience, but some performers can also use humour and indirection to introduce their songs. Glen Hansard is a folk-rock singer often defined by notions of authenticity due to his soulful delivery style, trademark beaten up acoustic guitar, and unaffected attitude after winning an Oscar for Best Song in 2006 (Leslie 2012; Gormely 2012; Mullin 2012). I suggest much of his authenticity is constructed during his charming inter-song stories and introductions. For example, the three recorded live versions of the song 'What Happens When the Heart Just Stops' (The Frames 2002, 2003, 2005) each feature a different story to introduce them. The stories either reinforce the heartfelt message of pain and loss in the song, or they paint a romantic picture of Hansard's childhood that misdirects from the lyrics of the song but enhances an awareness of 'what it is like to be him.' Both approaches give the audience space to construct authenticity and invite them to invest more attention to the meaning of the proceeding song, ascribing more authenticity as a result. Whether the stories are true or not seems unimportant in the construction of authenticity, as in Habermas' Theory of Communicative Action, the audience construct Hansard's intentions as sincere.

2. Second Person Authenticity

Moore's 'second person authenticity' happens when "a performance succeeds in conveying the impression to a listener that that listener's experience of life is being validated, that the music is 'telling it like it is' for them" (2002, p.220). Being part of a culture includes being exposed to certain prevalent works of art embedded in it and sharing an enjoyment of these works can lead to an authentic experience for those involved. This presents an anomaly in our pub singer's quest for authentication, as it suggests that Moore's 'second person authenticity' occurs when performing the most well-known songs he can muster; precisely the ones that make him feel inauthentic, for example, 'Wonderwall' and 'American Pie'. It is also an idea that runs against Heidegger's notion of inauthenticity associated with Being-in-the-world as the authenticity occurs in the experience of the listeners.

Discussion with Performers

I asked some fellow musicians, all who are professionally involved in original music but supplement their incomes by playing covers, how they felt about playing this kind of culturally entrenched material. An important notion that was raised was the importance of how well one is playing, regardless of the material. Certain songs have a stigma attached to them and deep-rooted associations from having been over-played badly by bad musicians and bad buskers who use this material because it is unchallenging for them and their listeners. A pub singer's audience brings this baggage into the room with them so performing a song like that risks inciting cynicism and being mired by these associations. The interviewees agreed that the solution was not to avoid the songs, but to play them with the highest level of energy and musicianship so the audience has a great time and respects the musicians. It was noted that playing the songs in one's own arrangement helps with feelings of authenticity, but the main point was that if a musician knows they are performing badly or not making enough effort, they will not feel authentic regardless of the audience or material.

The findings from this discussion are presented as a flow diagram in Figure 13.2, which demonstrates the performers' criteria to examine if they feel authentic during a performance.

Do I Feel Authentic?

3. Third Person Authenticity

Moore's 'third person authenticity' occurs when a performer "succeeds in conveying the impression of accurately representing the ideas of another, embedded within a tradition of performance" (2002, p.218). This mode of authenticity is achieved by the pub singer's choice of material. In Deena Weinstein's exploration of the history of rock covers, she presents postmodernist reasons for choosing material, which includes finding songs that express the performer's views better than anything they could write themselves (1997, p.146). These ideas can be examined by looking at how songs get into the pub singer's repertoire. Key factors are meaning and imagery in the lyrics. The song must fit with my social and political ideologies, for example with Bob Dylan and Radiohead; the lyrics need to have an intellectual or poetic edge that fit in with some romantic self-image, examples being songs

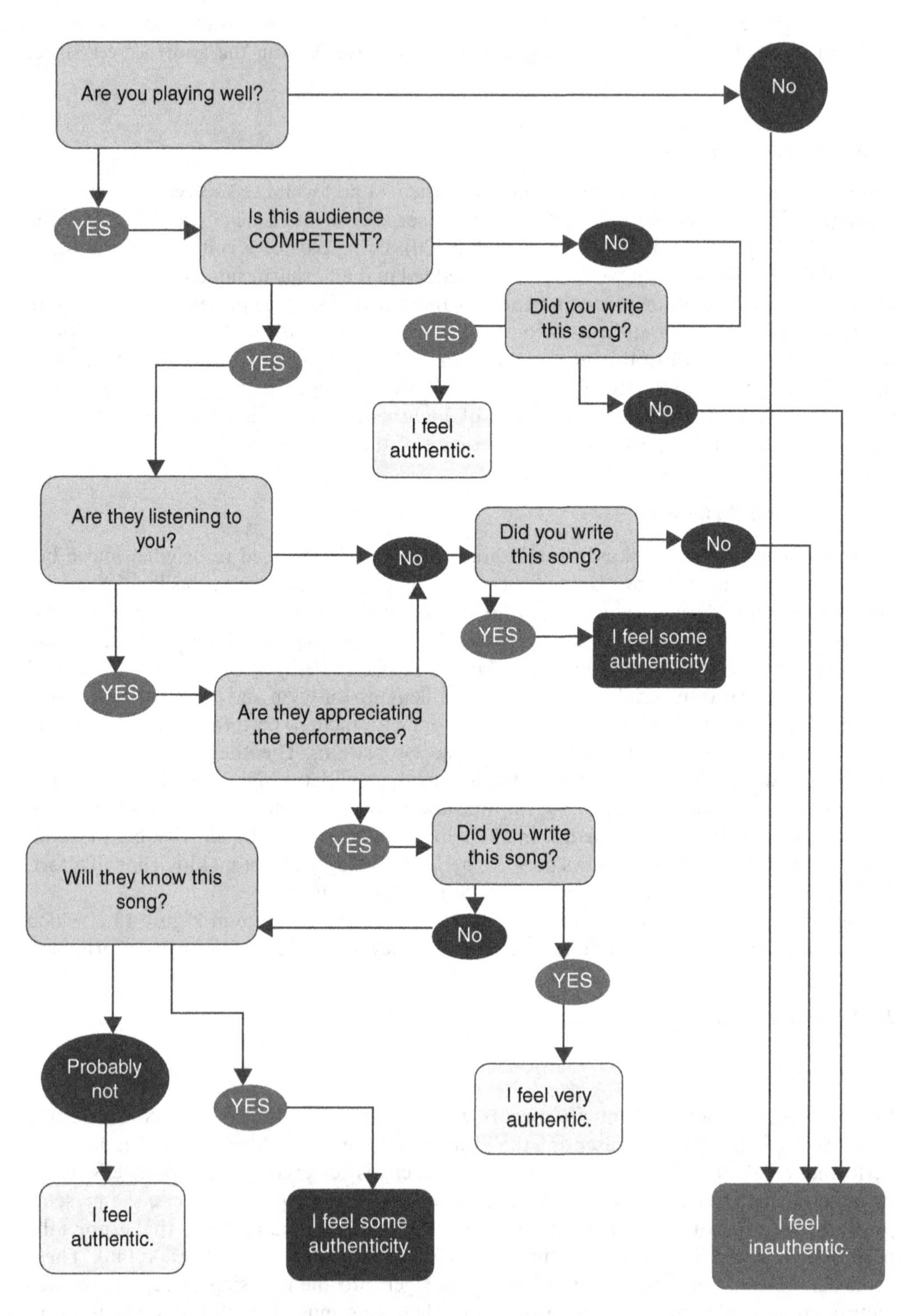

Figure 13.2 The performer's criteria for feeling authentic

by Tom Waits and Neil Finn; or they can be cleverly 'plain spoken' and beautifully reveal simple human truths, such as John Prine's and Steve Earle's songs; or they must narrate a dark story that I can get my performance teeth into, for example Terry Allen, The Band, and traditional folk songs. If a song fulfils some of these criteria then I feel that I can perform it authentically, but most importantly, I as a listener must have previously ascribed authenticity to the originator.

Typology of Me

> I usually use "I" as the narrator in my songs, but not all the "I's" are me; they're characters. It's theater.
>
> (Joni Mitchell in Diehl 2010)

There is a multiplicity of characters within the pub singer that are triggered by the pub singer's interpretation of different audience types and situations. In order to interrogate these relationships we will need a typology of Me. We are interested in how authenticity is ascribed differently to each:

- Me the Serious Singer Songwriter;
- Me the Intense, Challenging Rock Musician;
- Me the Non-Invasive Background Music Provider;
- Me the Upbeat Sing-along Entertainer;
- Me the Charming Speaker;
- Me the Digest of Interesting Songs – a music expert with impeccable taste trusted to seek out good songs and present them in an engaging way;
- Me the Accessible Rootsy Folk Singer;
- Me the Audience Member, assessing my own performance.

In our previous example in the wine bar, Me the Non-Invasive Background Music Provider was activated. Should that evening have progressed, and the suited gentlemen had become livelier and more interested in the music, Me the Digest of Interesting Songs may have taken over for a while before being usurped by Me the Sing-Along Entertainer if the gentlemen had had a few drinks and were getting rowdy.

Interview with an Audience

Having noted the theory and our typologies I carried out several informal interviews after a gig to better understand what a real audience uses to ascribe authenticity. It is recognised that only people who were engaged in the performance could realistically be canvassed, and of those, half were familiar. The interviews happened after a performance that included mainly original songs and fairy obscure covers, but with some well-known pop rock songs included for the purposes of this study. I asked them what they thought about my authenticity during the performance and paraphrase their response below.

> The performer is authentic because he doesn't play what everyone else plays.
>
> (Response 1)

Weisethaunet and Lindberg define 'Authenticity as Negation', evolved as "the antidote to the world in which popular music operates" (2010, p.472). It suggests I am authentic because of what I don't play, as measured against similar pub singers in the same scene and indulges a core myth of rock: "the image of the artist as a rebel" (Weisenthaunet and Lindberg 2010, p.472). It validates my artistic refusal to play well-known material. By giving this response the interviewees are ascribing authenticity to Me the Digest of Interesting Songs and Me the Intense, Challenging Rock Musician.

> We enjoy being introduced to new songs we wouldn't otherwise be exposed to.
>
> (Response 2)

This response also suggests respect and authenticity being ascribed to Me the Digest of Interesting Songs. Early discussions of authenticity in popular music were prompted by music collectors and archivists in the United States, like John Lomax and Harry Smith, seeking out music that they judged to be 'old time' or traditional and compiling collections of their field recordings (Barker and Taylor 2007, pp.31–57). Their taste was instrumental in defining what music could be called 'authentic'. In a similar way, Billy Bragg praises John Peel's taste, "The only criterion was that the music he played had to be challenging" (Plunkett 2012). In ascribing authenticity to the knowledge and taste of an Other, the audience can feel validated in return if they too enjoy the material. Weisethaunet and Lindberg paraphrase Zanes when they note the experience of authenticity as "an act of self-construction through an Other" (2010, p.477).

> The performer is authentic when he plays his own songs.
>
> (Response 3)

They are clearly ascribing authenticity to Me the Serious Singer Songwriter and reinforcing the observation that it is of benefit to verbally introduce the songs.

> We detect a darkness and melancholy in the songs that resonates with our world view.
>
> (Response 4)

Weisethaunet and Lindberg observe the power of the performer to hold up a mirror to the audience so that they recognise themselves in a song, and perhaps get "the feeling that the performer has been able to look into one's heart and has found the means to communicate who one *really* is" (2010, p.477). They term this 'Experiencing Authenticity' and in our case it seems to be ascribed to Me the Intense, Challenging Rock Musician.

> The inclusion of Man on the Moon by REM jarred in a set of otherwise more interesting material. It seemed less authentic.
>
> (Response 5)

It is interesting that 'Don't Look Back in Anger' was not mentioned, as that song was included to expect this response. Perhaps the performance of 'Man on the Moon' was simply not very good. Similarly, 'I Don't Like Mondays' was not mentioned, which was included as a radically different version to the original to see if it would provoke a response.

> We loved the idiosyncratic moments. The forgotten lyrics, mistakes, banter with the audience. They add personality and seem authentic.
>
> (Response 6)

This would seem to authenticate Me the Charming Speaker and reinforce the earlier discussion about song introductions. It is noted that Me the Charming Speaker is a performer type that is rarely available to our pub singer, because unless the mood is just right, he can come across as forced and inauthentic.

> It is very important for us to see the performer is enjoying himself and 'getting into it' for us to believe the performance is authentic
>
> (Response 7)

This response would seem to authenticate all of the performer Me's and reinforce the conclusions of the discussion with other performers included earlier, but it highlights the complexities of the pub singer/audience relationship in the construction of authenticity. It presents a circular reference where the pub singer is reliant on the audience to ascribe him authenticity so that he feels validated and can enjoy his performance, whilst the audience require the pub singer to be enjoying himself before they can ascribe him authenticity in the first place. Breaking out of this loop as quickly as possible at the beginning of a performance is an essential skill of the pub singer.

Conclusions

The remaining question, 'how does the performer *think* the audience is measuring his authenticity?' is a useful evaluation for our conclusions. It seems impossible to separate the answer from how he measures the audience's competence. Audiences are assessed using a flawed system derived from the performer's narrow experiences of existing in the rock music world, so certain audience types have a musical taste imposed upon them and fellow musicians are awarded disproportionate powers to alter his choices, driven by a warped assumption of their take on authenticity. The audience interviews above reveal an important difference in their actual attitudes: what the audience enjoys is different to what the performer thinks the audience wants. They should be validated just for being there. Any audience could contain keen music fans rebelling against the polish of the mainstream and seeking out the rawness and originality of a grass roots environment, just as much as it could contain people whose only exposure to music is mainstream television. Is it shallow to just play well-known songs for them? Probably, but the main issue is that it highlights a disrespect for the audience and consequently the performer. Heidegger's notions of authenticity do not allow for the influence of external forces, so if the performer is to focus on pleasing any part of the multiplicity of the audience, perhaps it should be the internalised – 'Me the Audience Member'. As in the phenomenology of Sartre, "to be true to oneself … is at the centre of his concept of 'freedom'" (in Weisethaunet and Lindberg 2010, p.466). We have also seen from the discussion with musicians that performers do feel most authentic playing original or obscure songs, but that it is not the most important factor; knowing that they are performing well is far more significant.

I have linked the pub singer's view of their own authenticity to how creative they are being. Sternberg cites the creativity "in the interaction between a person and the person's environment" (1999, p.83). This highlights a significant conclusion: if the relationship between pub singer and audience is vital to the ascription of authenticity, the relationship must be nurtured by spoken elements within the performance. This 'banter' creates a space for authenticity to be constructed by informing the audience of biographical details about the pub singer, encouraging them to recognise their own worldview being reflected, and giving them an insight into the song choices being made. It also sets up a dialogue in the room to better cope with the dreaded heckle "Play one we know!" should it rear its ugly head.

Part Three

Type 3. Coproduction Without Consent

Denial or Unknowing Collaboration

14 The Song of a Thousand Songs

Popular Music as Distributed Collaboration (Toast Theory, Part 1)

Robert Wilsmore

In Spinoza's posthumously published *Ethics* of 1677, the 7th Definition of the 'Second Part of Ethics of the Nature and Origin of the Mind' states:

> And if a number of individuals so concur in one action that together they are all cause of one effect, I consider them all, to that extent, as one singular thing.
>
> (Spinoza 1996, p.32)

The Song of a Thousand Songs

What if we perceive the whole of popular music as 'one singular thing', as one massive collaborative project? The combined work of composers, performers, producers, listeners, all types of musickers whose efforts have gone into constructing this vast phenomenon. We often speak of popular music as a thing with a life of its own, so let us give this natural predilection some substance, some value, rather than just as a passing metaphor. If the genius-author is still part of the construction of value then we have as genius–authors the joint authors of the pop project, where the many songs are no longer seen as an immense number of singularities but as an authored entity in itself. This magnum opus that encompasses the entirety of pop has a name, it is one song called 'The Song of a Thousand Songs' and is composed and performed by a group called The And.

In this chapter I put forward the concept of popular music as multiplicity, as the one and the many, and consider the general workings of the popular music community in its collaboration on the pop project. In the following chapter I set out to show that the ends of songs are not ends at all, or at least the former fixity of such endings has undergone a rapid dismantling in recent times, and put forward, as the main concept of Toast theory, the following axiom:

> **Musical phrases in which we perceive resemblance operate through the same modes of relationship regardless of whether they occur within one piece of music or in separate pieces of music.**

When we hear a theme develop in a symphony, we heap praise on an author. When we hear two similar themes in different songs, we heap criticism and shouts of 'plagiarism' upon one of the authors. Yet, in terms of what we *hear*, there may be no difference at all in these two scenarios.

DOI: 10.4324/9781351111959-17

'Toast'

> Thy cheeks are comely with rows of jewels, thy neck with chains of gold.
>
> The Song of Solomon (1:10)

Collaborators might normally be expected to be willing participants in a project, people who have signed up for a joint venture, but perhaps all of us who are involved in pop have signed up simply through our actions in that we have, in some way, 'contributed to pop music'. The very phrase suggests that we have played our part in constructing a joint *thing*. Through this lens we are all collaborators on the pop project, whether willingly or otherwise. For the purpose of this chapter I will use the terms 'pop' and 'popular music' freely without interest in any distinction, and I will take the position of allowing the pop project to be a collaborative venture into the construction of a singular entity, the many songwriters contributing to the on-going single that is popular music, the ever-shifting song that is 'The Song of a Thousand Songs'. This song is always number one in the charts and it is always in hibernation on a shelf, it is always under renovation, and it is always discarding old blooms that were once its brightest flowers. In times of posterity it is seen in full bling, bedecked with 'goldy-looking' chains, and in times of austerity it revels in the authenticity of unplugged soul searching. 'The Song of a Thousand Songs' is composed and performed by the group The And, a band that has itself seen a number of changes in its line up over the years. The And is, of course, the name of this conceptual group of collaborators, that is, everyone that contributes in whatever way to the construction of 'The Song of a Thousand Songs' is a member of the band (any relation to any other band named The And is entirely coincidental, although ironically, if there are bands out there with this name then their band The And is also part of The And).

Robert C. Hobbs in his essay 'Rewriting History: Artistic Collaboration since 1960' puts forward this proposition:

> I would like to propose that collaboration is, in essence, nothing more or less than influence positively perceived as part of an on-going cultural dialogue.
>
> (Hobbs in McCabe 1984, p.79)

The manifestations of these influences, songs, allow us to see the contributions to the project and how they are connected. If we stand further afield for a while and watch the musicking (Small 1998) happen then we will see how the capacity for influence that is ready and waiting in a receiver acts upon the received object and throws it out once more to be received by others with the capacity to respond to influence. The capacity for influence encompasses the readiness to receive, the desire to engage with what is received, and the ability to use the received in actively shaping a new output. It is the ability to be affected and for this affectation to have an effect, a manifest output. Collaboration is the effect of affect. As Deleuze writes on the Spinozian body, "a body affects other bodies, or is affected by other bodies; it is this capacity for affecting and being affected that also defines a body in its individuality" (Deleuze 1988, p.123). The concept of influence that has dominated this way of thinking in recent times is that of Richard Dawkins' meme as he describes it in his 1976 book *The Selfish Gene* (2016), and indeed his first example of the meme are 'tunes' (Dawkins 2016, p.249). The way in which the meme succeeds is outlined by Francis Heylighen through the processes of assimilation, retention, expression, and transmission (Heylighen 2009) and this aligns with the description of the capacity for influence drawn from Hobbs' statement. With this gene-like propagation now established as a mechanism for this cultural replicator we can consider this in light of collaboration. We are familiar with noting influence from one entity to a later other (through one author to another), so if we consider the collective endeavours of popular music as a manifestation of collaboration then we might see more clearly why the status of serious and non-serious music (in Adorno's terms) has evened out, for part of this discussion is with regard to the comparison of the thematic development found in the symphony (symphonic first movements in the Classical style in particular) with the thematic development (often termed 'the evolution') of popular music.

Vera John-Steiner and her collaborative partners note four patterns of collaboration that range from the widest form to the closest form: distributed, complementary, integrative, and family. These are not closed forms, this classification "is not a hierarchy" (John-Steiner 2000, p.197) and are represented within an un-segmented circle. She goes on to detail some of the characteristics, such as "Distributed collaboration is widespread. It takes place in casual settings and also in more organised contexts." The examples given are demonstrated through the conversation, as in the following:

> The participants in distributed collaborative groups are linked by similar interests. At times, their conversations may lead to new personal insights. When exchanges become heated or controversial, new groups may form to address issues in greater depth. Other groups splinter or dissolve. But out of such informal connections some lasting partnerships may be built.
>
> (John-Steiner 2000, p.198)

The role of the collaborator in this model is "informal and voluntary"; it is a contribution that is not necessarily agreed amongst partners other than that interested parties make contributions to the project. Their working method is characterised as "spontaneous and responsive". It might well be that contributors to the pop project may be calculated rather than spontaneous, or even with calculated spontaneity (as in the case of fabricating the casual in order to market a boy band, for instance), but it is normally an uncalculated expression and is largely impulsive in the manner with which it responds. It has a capacity for influence and a desire to influence others. A distributed model of collaboration then has voluntary, informal contributions to a project of shared interests. It describes the collaborative nature of the pop project at its macro level.

To what extent do we ignore the music prior to its latest ownership (its latest re-spinning) by one composer or another? We have not let go of the author. In popular music, as in many other art forms, the author has a central place in our value system that marks out songs as entities, setting them up against each other in a competition for originality and authenticity. In the last few decades the author has been killed off (Barthes 1977) and then resurrected again but as a diminished 'I', an I that knows its constituency to be the connection of many other preceding I's (Deleuze and Guattari 1987). But the priority of the artist–author is still a predominant hegemony in popular music that marks out the greatness of songs by the clear distance between an object and its *other*, and by the quantity of others that are said to have grown from the one. In this view the seminal song is precisely the singular origin, the seed, from which other entities are grown and despite variations their DNA is easily traceable back to the one (a platonic form perhaps? Possibly a song called 'This Song Sounds Like Another Song'?). And when we choose to make distinctions between songs that hold such similarity of DNA we play into the hands of Adorno's accusation of the pseudo-individual, the blinkered claim to originality that cannot see its own foolishness and naivety, we think there is choice but our spectrum is only between two very slightly different shades of grey.

Pop music need not be a manifestation of a platonic form, a perfect song, a song of songs from which earthly copies are but poor shadows. The view of pop put forward sits contra to the idea of a single form that proliferates slightly different (and imperfect) copies, or at least if that is one way it may be seen then we must acknowledge that it is not the only way because this view tags the song with the label of 'the poor copy'. Joseph Campbell's 1949 *The Hero with a Thousand Faces* explores the univocity of myth stories, all of which carry the same 17 stages but appear in different guises. In contrast 'The Song of a Thousand Songs' is not one entity disguised as a thousand different others, it is not one actor with many faces who has a 'real' face that it returns to, in that scenario the faces are characters but the 'real' actor has only one face. The actor is the one true legend behind the many myths, the monomyth. As Campbell states in the preface to the 1949 edition, "It is the purpose of the present book to uncover some of the truths disguised for us under the figures of religion and mythology" (Campbell 2008, p.xii). But the many pop songs are not disguising a truth, they are real in a Baudrillardian sense of the real where the simulacra is true (Baudrillard 1994), and many writers, from Auslander's *Liveness* (1999) on, have followed this line of enquiry with regard to recorded music. But there are qualities of the monomyth that hold steady for this project. Influence requires that something be received and that this experience be part of the next journey. The new song that is born from influence is presented to the community with its new wisdom. Campbell wrote that:

> The full round, the norm of the monomyth, requires that the hero shall now begin the labour of bringing the runes of wisdom, the Golden Fleece, or his sleeping princess, back into the kingdom of humanity, where the boon may redound to the renewing of the community, the nation, the planet, or the ten thousand worlds.
>
> (Campbell 2008, p.167)

With the renewal of the community comes a return to a start, but not the same start even though the surroundings are familiar. The theme may be similar, but it has been changed, varied, developed. The cycle repeats only inasmuch that we choose to notice the similarities and not the differences. Myths and melodies recycled but their return to form is not one of the shackled slave whose chains will only stretch so far. Pop is willing in its love of proximity, in its revisiting and renewal, the 'renewing of community' is one of its favourite tricks.

There are many monisms and ideas of the one and the many, like Campbell's hero, that might be brought into consideration and I mention them here only inasmuch to demonstrate that the discourse in this project has been running for thousands of years already and also to pre-empt criticism that other lines of enquiry should have been brought into the discussion. Indeed, they should but they might just have to wait for another occasion. Amongst the ideas that have influenced the approach here, the many that contribute to this 'one', include Plato's *Parmenides* (see Jowett's Plato 1931) where the dialogue moves around proofs for the existence or non-existence of the one. Alain Badiou in *Being and Event* (2007) starts from the position that 'the one is not', drawing on the Parmenides dialogue. Plotinus, in the third century BC, considers the will and the freewill of the one and Max Stirner (see Stirner 2014), in the 1840s, argued for personal autonomy in *The Ego and His Own*. Arthur Koestler's 1967 book *The Ghost in the Machine* put forward the 'holon' as something that is both whole and part. James Lovelock, in his 1979 book, proposed the living world of *Gaia* and this shares something of the quality of pop music in this study ('The Song of a Thousand Songs' has the quality of a living being). David Hume's 'bundle theory' and Wittgenstein's 'family resemblance' also play their part, and I have mentioned Dawkins' meme (Dawkins 2016) and will speak more on the multiplicity of Deleuze and Guattari's rhizomes. Artist and academic James Burrows recently proposed the concept of 'rhizo-memetic art' to capture the nature of art that operates with both meme-like and rhizomatic qualities, stating, "The non hierarchical nature of the rhizome coupled with a new understanding of its transitive, enacting qualities as mimetic generated a level playing field for all possible elements of meaning-making" (Burrows 2016, p.39). And, as if to evidence how rhizomes spread out on new lines of flight, in a post-graduate conference at York St John University, one presenter (a music production student) mis-spelled rhizomatics as 'Rhizomantics', the ensuing discussion concluded that this would be a great name for a band, in fact 'The Rhizomantics' might even have been a better name than 'The And' as the all-encompassing supergroup writers of 'The Song of a Thousand Songs'. A new variant had emerged that may or may not take hold. Musician Patrick Metzger (2016) coined the 5th, 3rd, 5th, 3rd alternating notes of a major triad as 'The Millennial Whoop' demonstrating Dawkins' meme in action (or Hobbs' influence, depending on how we view it). The news regularly produces new claims of pop theft. Take as an example Peter Robson's *Guardian* article 'The Songs Remain the Same' (Robson 2017); in this instance the trigger is with regard to the questioned authorship of an Ed Sheeran song that outlines the on-going problem caused by the separation of author (ownership) and thematic development drawing on analytical and mathematical probability in relation to musical authorship by experts such as himself and forensic musicologist Joe Bennett. In this article Bennett is noted as saying of similar phrases that they are likely to be a result of "some element of copying" and then adding, "Whether that's plagiarism becomes a legal question" (Robson 2017, p.5).

Copying, plagiarism, theft, influence, collaboration; there are so many perspectives by which we can view our musical authorships and there is so much yet to explore, but the two main ideas that I draw upon here are John-Steiner's distributed model of collaboration, and Deleuzoguattarian rhizomatics. The former concerns the human contributors, the subjects (the band The And), the latter the sound, the object ('The Song of a Thousand Songs'), although I shall be cautious not to play too heavily on the object/subject binary, and if Spinoza occasionally rears his head here then it might be as a reminder that there is only one substance, period, no Cartesian split in this anti-dialectical approach. Hence also why Dawkins' meme, when seen as a quanta, a singular thing, is not fore-fronted here even though its relationship to this discourse is obvious, although perhaps this chapter does attempt to continue the discussion posed by his question "I have said that a tune is one meme, but what is a symphony: how many memes is that? Is each movement one meme, each recognisable phrase of melody, each bar, each chord, or what?" (Dawkins 2016, p.253).

For the second part of this discussion (in the following chapter) it is the "recognisable phrase of melody" that will be the quanta, but the purpose will not be to count how many (we do that in later chapters on 'deproduction') but rather to view them as the continuous river, and that is perhaps best thought out through the concepts of the rhizome and the distributed contributors. I will, however, pick up directly on the symphonic reference made here, albeit in reference to a different Beethoven symphony.

If we are confident, post Barthes, in our constant recycling of ideas, of our playlists that overlap as songs drop off, songs remain and new songs added, then we can re-perceive the notion of popular music not as entities jostling for position amongst themselves, as we hear on radio chart shows, but as a collaborative network that may be viewed as an ever-shifting multiplicity. Shakespeare's 'All the world's a stage' reinterpreted as 'all the world's a song' or 'a symphony' or whatever we might wish to change the placeholder word 'stage' to, and then by continuation that 'all the men and women are but collaborators'.

Deleuze and Guattari shouted at us in *A Thousand Plateaus* that "RHIZOMATICS=POP ANALYSIS"(1987, p.26) and through the re-envisaging of culture as the ever connecting and disconnecting rhizome in place of the arborescent (tree-like) fixity of existing hierarchies, we might see that it is not simply a hypothetical discussion to view the pop project as the workings of influence, even that influence *is* the rhizome (rather than just the mechanism), and so it is an actuality that we jointly author this large work. The pop project is an authored magnum opus, just as Beethoven's 5th Symphony is an authored magnum opus. Hence, we are not surprised by our omnivorous appetite for both pop and classical music, and to give them a greater sense of equality than we see in Adorno's 'serious and non-serious' distinctions. Peterson and Simkus (1992) noted the move towards cultural omnivorousness in the late twentieth century. Peterson and Kern then stated that "We speculate that this shift from snob to omnivore relates to status-group politics influenced by changes in social structure, values, art-world dynamics, and generational conflict" (1996) and I hope that this study might make a small contribution to the workings of cultural omnivorousness. Many of us have lived through a time where the status of pop and classical has altered, evened out, not dumbed down. It is not simply a reluctant shrug of the shoulders and a sigh that relativism has levelled the playing field, however true it might be that there are no truths and hence no way of proving classical above pop. Relativism need not equate to passivism. There are explanations for this shift and I hope to provide some here.

But if we stay with the concept of the multiplicity, in this respect we might see a song and its predecessors (and its successors) as part of the same piece, it is not an *inter-textual* construction but an *intra-textual* one. Just as Adorno might favour the thematic development of a symphonic theme as central to the integrity of the work, and Deleuze the Internal difference (the postmodernist Deleuze is a modernist at heart when it comes to music), so we might see the lines of flight from song to song as we see the relationship between the opening four notes of Beethoven's 5th Symphony and the following four notes (which will be worked through in the next chapter). They share a pattern but are not the same, they are a repetition but are also unique; they know each other, acknowledge each other and depend on each other, and are part of the same project.

As Nick Nesbitt (in Buchanon and Swiboda 2004) notes, in working through a reconciliation between Adorno's dialectic and the Deleuzian anti-dialectic (a co-existence of both states in an apparent binary) it is possible to use these ideas to make sense of pop in a way in which it was perhaps not intended. Indeed, in Adorno's case the binary of serious and non-serious was drawn up specifically with the intention of setting apart classical from pop. It is not that I wish to deliberately pervert and reverse the concept, Ranciere in his essay on the *Misadventure of Critical Thought* notes the pointlessness of such reversals writing "I do not want to add another twist to the reversals that forever maintain the same machinery" (Ranciere 2011). In this case it is a response to the observation of the levelling of status of pop and classical that brings me to view the pop project as sharing the qualities of the thematic development of the classical technique and the qualities of Vera John-Steiner's distributed model of collaboration. It is not a will to prove either Adorno wrong or Deleuze right (their stances are not so simply one sided anyway and are historically situated in the discourses of their times) but rather to show that their concepts are applicable towards pop in a positive light in a way in which they themselves did not do, for whatever reason, at their time of writing.

If we follow this line of enquiry and see the pop project as one symphonic masterpiece (if we now take the 'master' as the collaborative many, the distributed contributors, and not the solitary genius) we might also see the sole authored symphonic masterpiece as merely one fragment, one variant of a developing theme that plays its part in the collaborative effort of the classical project. I might have begun this article with the question 'What if we perceive the whole of music as one massive collaborative project?' but the exposing of how value judgements are constructed across genres is worthy of the specification here and it is the disintegration of the classical and pop divide that we witness in our society in our omnivorous habits that leads me to enquire on this shift. And to reiterate, this is not a rage against the old 'serious, non-serious' argument but observations surrounding its disintegration, or rather its integration.

Stefani (in Middleton 1990) notes that there are poles in our reception of music from popular competency to high competency, and we might be able to pinpoint each individual on the scale. Some musickers will locate themselves firmly in one position, Adorno clearly only occupying the position of 'high competency' as a listener, the listener that can grasp the internal logic of any work however unfamiliar it might be. The listener that turns off quickly with the accusation of 'that's rubbish' as soon as a new piece of music falls outside their reference framework that has similarity as its primacy is pinned down at the bottom of the scale as only having 'popular competency'. But it has been increasing less like this for decades now, many are as content to deal with the complexity of the internal workings of a modernist composition as they are engaging in pop.

Neither is this an analogy that suggests that we are happy at times to wear our Sunday best but then at other times to slob out in our PJs; that analogy merely states that we are still aware of the comparative values and the hierarchy is still in place. Catwalk fashion and high street fashion. That is not it. We engage with pop with the same level of sophistication as we engage with the most complex works that the modernists may have thrown at us. The accusation will be that although we say we are listening to complex work we are not really understanding it because if we did, we would not be able to listen to pop with the same level of value attached. But the levelling of pop and classical is genuine. Once we recalibrate the perceived distance between the classical composer–genius and the pop-makers as a collective then it begins to clarify as to why classical and pop have levelled out in recent times. Hobbs' notion of 'influence as collaboration' then is key to this shift. We need to make a leap, in fact it is evident to many of us that we already have made this leap in our usage of music, and that is that we need to compare the classical composer to the pop collective, the symphony to the playlist, genius to scenius. This is not to say that Beethoven is worth a thousand pop artists, that Mozart equates to a dozen Michael Jacksons, that is also a misunderstanding of the value in place here. George Martin observed the shift that was occurring during the latter part of the century, as he so poignantly puts it in his 1979 book (with the grammatically dubious title) *All You Need Is Ears*:

> To be fair, most composers writing contemporary 'classical music' are in a cleft stick. They can't use the styles which have already been evolved, because then they're accused of being romantic, or sensuous, or derivative. So the only way to go is to write new sounds – and remember, even the twelve-tone scale is old-fashioned now, and almost regarded as romantic in itself. The result is that the modern 'classical' composer either writes stuff that most audiences can't stand, or reverts to symphonies that could have been written by Brahms. And what's the point of that? So 'classical music' becomes a one-way street; and that's where pop music has come into its own, because it can be truly creative.
>
> (Martin and Hornsby 1979, p.31)

The great creative force in music in our time has been popular music, but like all such blooms of creativity, it is likely to end at some point and itself end up in a position similar to that which Martin describes above. How it will end is only something we can speculate on, which of course we do in the final part on 'deproduction'.

Das Leibe Farbe

In an argument in *Deleuze and Music* (Buchanon and Swiboda 2004), Nick Nesbitt finds an inter-subjectivity in Coltrane's work that shows jazz to meet the criteria for 'serious' music that Adorno says jazz lacks and hence proves jazz to be 'serious' using Adorno's own criteria. Adorno's internal criteria for serious music is characterised as "Every detail derives its musical sense from the concrete totality of the piece which, in turn, consists of the life relationship of the details and never of a mere enforcement of a musical scheme" (Adorno 1941). Adorno ignores the observation that the rules of internal thematic development are also externally enforced schema. The rule of thematic development states that successive manifestations of the theme be derivations of the theme; and this is the external, pre-determined rule that governs this particular compositional ideology. Nesbitt notes of Deleuze's concepts that, with regard to this working out of Coltrane's internal integrity of his music:

> It is equally the truth of a Deleuzian notion of internal (musical) difference, as it constructs for us the selfsame musical entities that (consciously) refer only to themselves to the working through of their (musical, existential) problems, and in this utter, windowless interiority, resonate outwards across, time and space, culture and race to express fully what a [living] body can do.
>
> (Buchanon and Swiboda 2004, p.72)

The 'windowless' room that Nesbitt describes in relation to Boulez is a space that attempts to leave all relational associations outside and allow the purity of internal logic to do its thing. It asks for totality, and not surprisingly culminated in 'total serialism':

> Boulez strives to purify his musical plane of all negativity ('no denials to confront us') so that he may proceed univocally. The serial world Boulez constructs is quite literally univocal, an apartheid-like space [...] in which no dissenting voices can deny the validity of the Serial musical procedure.
>
> (2004, p.66)

It is not necessarily a misguided notion that this univocity in music occurs, that the internal logic of a closed room that allows no external connection is a bad thing. It strives for universality but within a finite universe that (ought to) accept the multiverse, the co-existence of other universes, but chooses to focus on itself (it has no access to other universes). It is an ambitious project, to be immersed in this pure world would be a delight, after all ignorance is bliss, and we can of course return to engage ethically in the world post immersion. But it is a choice to claim this to be the goal of music, a local decision that chooses to place one criteria above another; there need be no fixity here "for there is nothing either good or bad but thinking makes it so" (*Hamlet* 2:ii), or even, from the same century as *Hamlet*, Spinoza's observation that "perfection and imperfection, therefore, are only modes of thinking" (Spinoza 1996, p.115). The assumptions here are with regard to whether we label according to content or exclusion. In phenomenological terms we see leaves as being green and so we call them green. The leaf itself, were it able to engage in critical debate, and that it is a leaf that engages with itself on a more internal level than humans do (but we could do if we wished), might complain that it is anything but green, it appears that way because it is the colour that it rejects from the light shining on it. We might place value on a nightclub according to whom it keeps out, whom it excludes. The club then is exclusive.

The windowless room is likewise exclusive, its position defined by whom it keeps out, and whom it keeps out are those that refuse to disconnect with on outside, who refute the demarcation that creates an outside. Let us be *inclusive* and see all pop music, in all its wonderful colours, as being connected and as a constantly becoming entity, let us acknowledge that we are listening to one song that grows offshoots, that breaks and then starts up on new lines of flights, that connects in an ever-shifting map, not a fixed trace, not a fixed hierarchical lineage but a timeless middle.

15 Removing Non-sonic Signifiers from Endings (Toast Theory, Part 2)

Robert Wilsmore

In the Preface to the 'Fourth Part of The Ethics of Human Bondage, or the Powers of the Affects', Spinoza states:

> On the other hand, he will call it perfect as soon as he sees that the work has been carried through to the end which the author had decided to give it. But if someone sees a work whose like he has never seen, and does not know the mind of its maker, he will, of course, not be able to know if that work is perfect or imperfect.
>
> (Spinoza 1996, p.114)

So, what is an end, and who decides when the music is finished? In this chapter I discuss how musical phrases ('tunes') can be perceived in relation to each other and propose that the observed borders of a traditional view of musical entities is no longer as fixed as it once was. In order to do this, I will consider two case studies: One where two similar phrases are accepted as an internally validated thematic development within a single work, and the second where two similar phrases are accepted as being from different works. In this way we can then ask the questions as to how the entities (the pieces of music) are made separate. The two case studies are:

1. Thematic unity with a single entity: the opening of Beethoven's 5th Symphony, in particular the opening theme (A1) and its first variation (A2);
2. Thematic unity across multiple entities: a phrase in song 1 (S1) and a similar phrase in song 2 (S2).

There are two parts under consideration, first the ways in which themes may be seen in relation to each other, and second what divides and what unites these themes when they are internal (within one work) or in separate works (not within one work). Most of the focus will be on establishing the different ways in which themes may be seen to relate through the first case study. The second case study is shorter in as much as it is then tested against the relationship models established.

Case Study 1

Let us concentrate on the connection drawn between the proceedings within Beethoven's 5th Symphony. If we take the opening theme (A1) as the original theme we hear the next phrase (A2) in relation to this (Figure 15.1).

We can note similarities in the phrases, such as, for example, the following:

- The rhythm 'short, short, short, long';
- The 'shorts' are a repeated note (pitch);

DOI: 10.4324/9781351111959-18

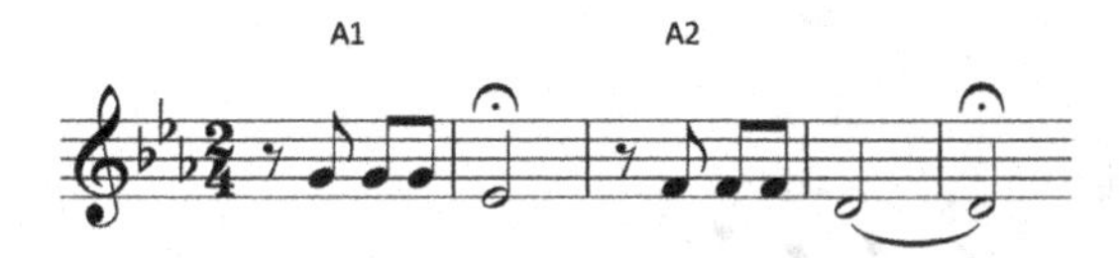

Figure 15.1 Two phrases within 'one' piece

- They are the same duration (quavers);
- The interval from short notes to long is a 3rd.

And we can note some differences, for example:

- A2 comes *after* A1 (it is not before or at the same time);
- The phrase A2 starts a tone down on the F and not the G;
- The intervalic descent is a minor 3rd and not a major 3rd.

Putting aside the larger discussion with regards to the entity of the symphony as a whole, we might concentrate locally on the theme here in relation to being an entity and there are a number of competing models for how we might view the proceedings of Beethoven's 5th. To begin we will look at two simple approaches:

Mode 1. The theme (A) is a *singular* entity that is observed to develop over time;
Mode 2. The themes (A1, A2) are *multiple* entities that are related to each other.

In the first mode we might analogise the theme as a person; when looking through a photo album we might see the first photo of the person as a baby and the second as a two-year-old toddler. We understand that this is the same person (the same entity) but that they have changed over time. In this way we see A1 as the baby and A2 as the same person that has changed over time. We may see contiguity on the score where both phrases are there at the same time on the page and give an illusion of being separate, but in the linear world of sound we hear A1 start and end, then it changes and is A2, then A3, and so on. In this way we look at A1 as we watch a child grow up; A1 aged one year, then A2 at age two years, three years, and so on. It is one entity that we see develop over time.

In mode 2 we hear A1 as a single entity, then we hear a new single entity A2 that has been influenced by, or spawned from, the previous entity (separate memes perhaps). In this model it is not one entity that changes over time but separate entities, multiples, that are related, that are mutations, variations of the original. In this second model, where the linearity of the music suggests the primacy of the first theme, A1 is the 'parent' from which others grow rather than the infant that is seen to develop. A1 is the mother to A2, and A2 the mother of A3, or maybe A1 is the parent of both A2 and A3, which are siblings, a mode that is akin to Wittgenstein's 'family resemblances' perhaps (see Wittgenstein 1953).

With both modes in mind then we can see a fundamental duality in the phenomenology of thematic development; the single child that develops, and the parent and their children. Both of which, although only analogies, are equally valid models by which we see the relationships between musical phrases. Our next step is to consider how these possible perceptions work with regard to noticeable similar musical phrases in works that are *not* considered to be within the 'one' piece of music.

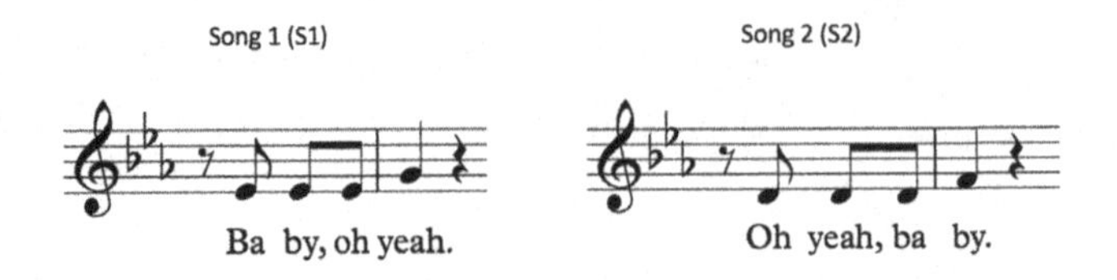

Figure 15.2 Two phrases in 'separate' pieces

Case Study 2

The theme in song 1 is labelled S1 and the theme in song 2 is labelled S2. They are (hypothetically) by different composers, they are in different songs, some listeners think that they sound similar. I am not using 'real' examples here, there is no need, and besides I am not keen to suffer the painful legal actions that might follow if I did. So, I will use instead the following made up, but quite plausible, examples (Figure 15.2).

Here we take two phrases S1 and S2 and view them through the same modes as we did for A1 and A2:

> Mode 1. The theme (S) is a singular entity that is observed to develop (over time);
> Mode 2. The themes (S1, S2) are multiple entities that are related to each other.

We can observe that:

- S1 can be seen as one entity that, on hearing S2, we note has changed over time;
- S1 and S2 are separate entities, but we perceive that they are closely related.

And it should be evident that many of the same similarities apply. We can note similarities in the phrases:

- The rhythm 'short, short, short, long';
- The 'shorts' are a repeated note (pitch);
- They are the same duration (quavers);
- The interval from short notes to long is a 3rd.

And we can note differences:

- S2 comes *after* S1 (it is not at the same time, at least not on the page here);
- The phrase S2 *ends* on the F and not the G;
- The intervallic ascent in S2 is a minor 3rd and not a major 3rd.

It might be obvious in this example given that the 'song' phrases are deliberately drawn from Beethoven's theme, and hence are bound to end in the same result. But that only goes to emphasise the point, S1 and S2 might well (and very probably do somewhere in the millions of 'actual' songs) exist. Both modes of viewing the thematic relationships apply equally to both case studies and hence from this we can put forward the axiom stated at the beginning of Toast theory:

> **Musical phrases in which we perceive resemblance operate through the same modes of relationship regardless of whether they occur within one piece of music or in separate pieces of music.**

So, the questions then start to become obvious when we ask 'what are the differences?' Let us pick out some of the obvious objections and discuss their merits. By objections I refer to those that draw such distinctions and privilege the coherence and totality of individual musical works over the joining of all works into one thing.

It might be said (in a derogatory way) of *similar* themes in *different* compositions by *different* composers that:

- They are not in the same 'piece of music';
- They are not by the same composer;
- There is no 'fixed' gap between when one phrase is heard and the next (could be months or years between hearing them);
- There is lots of other 'stuff' (other songs etc.) sounded between these themes;
- They are not *meant* to be together, but in the 'singular' work they are meant to be together;
- The second artist is a plagiarist, they stole it from the first artist.

Which out of these objections is an objection on musical (sonic, sound only) grounds? I argue that the answer is 'none'. Do we have to accept the objections that are set out here and adhere to their legitimate claim with regard to how they separate? We might simply ask then, why does it matter that they are not within the same composition? We still hear the relationship, the development, etc. The singularity of a piece is only defined by, for example, the end of a written score, the end of the record, the stream of the playlist etc.; they are *signifiers* of endings only, we do not have to accept them as concrete closures of entities.

That the placing of the theme in a fixed temporal relationship to its 'original' has priority over a less fixed and variable temporal relationship is but a preference of the listener or the composer, not a necessity (I refer to the fixed quaver rest between A1 and A2 in comparison to the unknown time and order of S1 and S2), and certainly not a necessity having been through a modernist age where a piece identified as an entity of itself can be so open to chance, aleatoric operations, improvisations etc. that such fixity will not occur even within that singularity. Even within a first movement of a classical symphony we are likely to find 'lots of other stuff' between the sounding of a theme and a subsequent development. But will A2 always come after A1 in Beethoven's symphony? No, it won't, A2 will at some point be followed later by another rendition of A1. It may be further away than a quaver but there will be a sequence that is A2 A1, just as in the case of the two songs we might hear S2 before hearing S1. The separations then appear to be only through two main operations:

1. The primacy of the composer (the name, the person, the owner) forces closure ensuring validity is internal;
2. Ends are given meaning through *signifiers* (the end of the score, the end of the recording etc.) but these signifiers need not have *veto* that enforces closure.

There are sonic signifiers that may lay claim to closure (a perfect cadence followed by silence and then applause perhaps) and I have willingly left them out of this discourse for the moment. That is not to deny that a perfect cadence followed by silence signifies an ending, but to include these at this point would put a screen up in front of other issues that do not deserve such protection. There are more nuanced investigations to come that will include cadences and silence but for now I have put them to one side for good reason.

Territory (Mode 3)

To return to where I began. I do not set out to break some classical and pop divide, that for many, maybe most, of us happened a long time ago. This is to explore just one aspect of *how* that divide broke. 'The Song of a Thousand Songs', if we are prepared to allow such a manifestation as co-existing alongside our grounding in the 'small refrain', happily accepts the validation of internal working for there is no external, we have panned out to the limits of our known universe. As listeners we hear familiar themes develop as we listen to our radios, often we are not even sure who the artist or the composers are, we are just listening and it all seems like one evolving body. DJs have become both angels and demons as far as Toast theory is concerned; Angels because they segue with as smooth a transition as possible from one track to the next, but demons when they stop to announce, "that song was by *x*, and the next song is by *y*".

I will propose a further mode by which we might view the relationship of themes just as an example of other ways that similar themes can be heard, and in keeping with the main philosophies here the third model concerns the working of the rhizome. Mode 3 is the Deleuzoguattarian multiplicity within which we no longer see one thing as representing another thing, but rather that a territory is continually deterritorialised and reterritorialised. The concept of the rhizome (and the rhizomatic) put forward by Deleuze and Guattari in *A Thousand Plateaus* proposes a rethink of our connective processes; that we stop thinking about trees with their singularity and their fixed hierarchies but instead consider the rhizome that connects, disconnects, reconnects and is constantly moving along new lines of flight establishing a logic of The And. One consequence of this re-conception is the removal of representation through an *a-signifying rupture*. An example is the wasp and the orchid where we no longer see the orchid as mimicking the wasp but as *becoming* the wasp. This process concerns the territory of the code and the continual duality within how one is deterritorialised and reterritorialised by another *other*. Deleuze and Guattari occasionally offer up direct comments on music. In *What Is Philosophy* (2011 [1991]), their last collaboration, they give an example, example 13 (2011, p.190), of how compositional processes have moved from classical times to modernist times. It misses in its aim in as much as it refers to individuals (Liszt, Wagner, Mahler, Boulez, etc.), which are clearly held up as the greats, as totalities, as 'trees', as authors and all that authorship owns (and trees are a big problem for Deleuze and Guattari), yet the notion of the small refrain (as within one piece of music) and the great refrain as an operational duality within the territorial system binds it as a multiplicity. Perhaps unfortunately (for Toast) the great refrain is encapsulated in a manifestation of it in Mahler's *Das Lied von der Erde* rather than being permitted to exist as the connections across refrains in multiple works. They state:

> All tunes, all the little framing or framed refrains – childish, domestic, professional, national, territorial – are swept up in the great refrain, a powerful song of the earth – the deterritorialised – which arises in Mahler, Berg, Bartok. [....] The great refrain arises as we distance ourselves from the house, even if this is in order to return, since no one will recognise us anymore when we come back.
>
> (Deleuze and Guattari 2011, p.190)

Some 25 years previously, in 1968, in *Difference and Repetition* (2010 [1968]), Deleuze wrote:

> For the Large and the Small are not naturally said of the One, but first and foremost of difference. The question arises, therefore, how far the difference must extend – how large? how small? – in order to remain within the limits of the concept.
>
> (Deleuze 2010, p.38)

In their anti-dialectical approach, they recognise the co-existing duality of the closing and opening that occurs in composition:

> In fact, the most important musical phenomenon that appears (as the sonorous compounds of sensation become more complex) is that their closure or shutting off (through the joining of their frames, their sections) is accompanied by a possibility of opening onto an ever more limitless plane of composition.
>
> (Deleuze and Guattari 2011, p190)

Deleuze and Guattari offer further concepts as well as that of the rhizomatic (and philosophy is "the art of forming, inventing, and fabricating concepts" (2011, p.2)) and Bodies without Organs (BwOs) and Planes of Consistency are fluid spaces and relate to the phrase here that the closing and opening move into a "limitless plane of composition". Here then is a third aspect of the thematic relationship of A1 and A2, and S1 and S2:

> Mode 3. An asignifying multiplicity where the territory is constantly deterritorialised and reterritorialised.

In this way we might view A1 as having a particular territory, that of the small refrain, and that at the sounding of A2 then A1's territory is deterritorialised by A2, and that A1 and A2 might deterritorialise and reterritorialise as a great refrain (or at least a slightly greater refrain as we 'distance ourselves from the house'). The complexity here is that we should not see this as a 'land grab' by one party from another, but rather that there is a mutual 'becoming'. A2 is a becoming A1 and A1 a becoming A2 and as such we should no longer be labelling entities as A1 and A2 as this assumes separation. There is a significant philosophical shift here, it asks us to doubt our assumptions of things as distinct entities that operate within a representational system. The orchid does not represent wasp, it is wasp. Similarly, then A2 does not represent 'the' A1 in this view, it is A1. But it is not A1 now as a toddler, where once it was a baby, here is the apparent dichotomy, toddler and baby now co-exist as a duality not as a dialectic or a binary, but neither are they the mother and daughter model, these modes of thinking are not compatible with this third mode. They are within the multiplicity, the rhizome, they are the one *and* the many. At this point once again the analogy to people breaks down and becomes non-sensible because they are no longer entities in themselves. Hence the first two models, 1 (the developing child) and 2 (the parent and child), no longer operate when viewed as the multiplicity that connects and maps and finds new lines of flight.

So, to recap these ways in which to view theme A/S and A1/S1 and A2/S2:

> Mode 1. That the theme (A or S) is a singular entity that is observed to develop over time;
> Mode 2. That the themes (A1 or S1, A2 or S2) are multiple entities that are related to each other;
> Mode 3. The theme(s) operate in an asignifying multiplicity where they are continually deterritorialised and reterritorialised.

The Collaborators

The focus on the separations and connections of thematically similar tunes detailed above has necessarily removed the author, and if this project on collaboration is to have credence then the 'people' need to be brought back into the discussion. As Vera John-Steiner's model noted, collaborators in a distributive model contribute because they are interested in the same project that others are interested in. She also notes that arguments arise between contributors where "exchanges become heated or controversial", just as songwriters get cross (and get legal) when they think another writer has copied them. Vera John-Steiner might as well be speaking directly about pop artists and bands when she writes that "new groups may form [...] other groups splinter or dissolve" (2000, p.198). I have said nothing about copyright in this discussion, this is not about what is legal or lawful and I have avoided direct comparisons of known tunes and the "heated exchanges" of actual people over authorship and ownership. If John-Steiner's collaborative model is to be of use then the observation must fit with the model. The contributions happen because of similar interests, and they are made in the way that Hobbs describes collaboration as "influence positively perceived as part of an on-going cultural dialogue" (albeit it might work better if the word 'positive' be removed to leave it neutral because of all the heated shouts of 'plagiarist!' that also impose the negative on it). But we also hear the voices that say that it is just the natural evolution of music, that it is just the continual symphony of sound, that there is nothing new anyway ("we've always used the same three chords"), that there must be a finite number of (likeable) tunes. That groups form, splinter, and dissolve is true of the artists themselves and also of the fans, the listeners, and users of the music (everyone involved from production to consumption, aka 'the musickers'). If we can separate the ownership issues that dominate the news headlines from the way we all use and contribute to popular music, then the Gaia-like model makes a great deal of sense. I have written many words (the result of many years of arguing it through) to labour the points about connection, but it might be that the most important aspect is that this thing now has a name; we always knew what it was, that it existed, it's the evolution of pop, it's rock n roll, but we might now refer to it by name, 'The Song of a Thousand Songs', and by that we will still all know what it means, it's just that we've given it a name. In this way we might understand that this thing contains both the recognition that there are multiple contributors *and* the acknowledgement that these contributions are made to one intra-textual project, not a mutually exclusivity but a multiplicitous reconciliation of the two.

Long Live Rock n Arctic Roll

In terms of the pop project it is normally seen as multiple entities that are related to each other rather than being mode 1, a single entity that is observed over time. We might be less familiar with equating songs to the deterritorialisation and reterritorialisation process of an a-signifying rupture, but pop as an entity is entirely familiar to us. To give an example at the time of first presenting this 'paper' (at the 2014 Leeds College of Music FIMPAC conference), the 2014 Brit Awards had recently been held on 19 February 2014. Alex Turner of the Arctic Monkeys on collecting their final award of the night said this:

> That Rock n Roll, it just won't go away. It might hibernate from time to time, sink back into the swamp. I think the cyclical nature of the universe in which it exists demands that it adheres to some of its rules. But it's always waiting there just around the corner ready to make its way back through the sludge and smash through the glass ceiling

> looking better than ever. Yeah, that Rock n Roll, it seems like it's fading away sometimes but it will never die. And there's nothing you can do about it.
>
> (Turner 2014)

And then, as if to prove how the capacity for influence is re-spun, Turner carefully and deliberately 'drops the mic' to the floor. As it hits the ground it echoes of Hendrix and burning guitars, of The Who and smashed amps, of mythical televisions through mythical hotel windows, and the fire that leaves the smoke on the water.

The And

Like the Arctic Monkeys, The And have also received many awards, in fact they have received every award. When supergroup The And first came to prominence in the early '60s (century unidentified) they were already dogged by accusations of unoriginality and pseudo-individualism. And because they only ever produced the one song, 'The Song of a Thousand Songs', they were soon labelled as 'one hit wonders'. Some critics jibed that the song title was quantitatively misleading with quips of "don't you mean a billion songs?" but the group just cited that the 'thousand' in the title was artistic licence that captured the idea of 'one and many'. Other critics went the other way and suggested that one thousand different songs was way over the top, that actually all songs were basically the same and so it should be called 'The Song of the One Solitary Stupid Song'. The band shrugged off the inclusion of the word 'stupid' as just a bullying tactic by the press but expressed that the 'Song of the One song' was kind of what they meant but not quite in the same way that the critics meant it. It wasn't trying to fool anyone by only pretending to be 'different when really it was exactly the same'. That wasn't what they meant at all.

The End?

A summary:

1. One song can be heard as the thematic development of another song just as the theme of a classical symphony is heard to develop.
2. Popular music when seen as a singular entity that continually connects, disconnects, and moves along new lines of flight is rhizomatic.
3. The manner in which musickers contribute to popular music is that of the distributed model of collaboration described by Vera John-Steiner.
4. The workings of this distributed collaboration is "influence positively perceived in an ongoing cultural dialogue" as described by Robert C. Hobbs.
5. The multiplicity that is 'The Song of a Thousand Songs' is intra-subjectively connected through the manifestations of influence between members of The And.
6. The And would like to thank their management, their sponsors, and everyone they've ever met.

16 The Ancient Art of Remixing

Robert Wilsmore

Background

This talk was written in 2004 and presented at a musicology forum at Leeds College of Music (now Leeds Conservatoire) later that year. Ten years prior to the first talk on the concept of 'The Song of a Thousand Songs', it details how a composer struggling with the concept of originality at the turn of the century sought to reconcile creativity with the use of existing material in the creative processes of composition. It is included in this book partly for its own discussion on tracing a 'new' work backwards in time with regards to its use of the 'pre-existing' (rather like the hypothetical first song 'This Song Sounds Like Another Song'), and partly because it is clearly a forerunner of Toast theory as a concept (see also Wilsmore 2010). It is somewhat naive and had already been surpassed as an essay on musical borrowing even at the time of its presentation, but the collapsing of the years that connect the twelfth and twenty-first centuries is a significant part of how Toast theory was emerging at the time.

Introduction

In the summer of 2003 Lakeside Arts Nottingham commissioned an orchestral work that was premiered by Viva: The Orchestra of the East Midlands on 22 April 2004, at the Djanogly Hall, Nottingham. The programme contained the works Bach *Orchestral Suite No.2*, Haydn *Cello Concerto in D*, Mark Anthony Turnage's *Kai* and Haydn *Symphony No.44* ('Trauer'). I was asked to respond to the programme in my piece. As a composer I struggle to put one note in front of another, and the notion of being an 'author' or 'composer' is for me a difficult one. Hence the piece that conductor Nicholas Kok and the Viva Orchestra received was a score that re-ordered and repeated some of the bars from the Presto from Haydn's Symphony No. 44. No notes were changed or were cleverly rewritten, and the concert programme reads '*Selection 44* by Haydn remixed by Robert Wilsmore'. The title relates to the number of the symphony, to the technique of choosing a selection of bars to re-order, and from contemporary associations with the words 'selection' and 'remix' (for example, Aphex Twin's *26 Remixes for Cash* and, more specifically, Squarepusher's album *Selection Sixteen*). I chose to use the terms 'remix' and 'selection' to associate with, and perhaps to be authenticated by, the dance music movement, which was the closest influence on the work, even given the obvious connections to minimalism in the use of repetitious material played by classical forces.

In 'On', the what's on guide for Lakeside Arts, the description read "Robert Wilsmore remixes the Presto from Symphony 44 to give a unique and very contemporary response to this work." It is pleasing to be acknowledged not just as a contemporary composer but as a *very* contemporary composer, clearly one who could offer a 'unique response' to the commission. However, it is of course foolish to believe the hype, and following a chance to

DOI: 10.4324/9781351111959-19

play the work in a research seminar at Leeds College of Music, a colleague remarked "of course you realise that this work comes from a long-standing tradition". As a *very* contemporary composer who has pretensions to be at the radical cutting edge of music, being placed in the comfort zone of a tradition is a little disconcerting. As always, my view of the event at the time was that it was far more experimental than it now appears in retrospect.

The use of existing musical material in new works has been in existence at least as far back as records go, and my use of the word 'remix' as I have said, makes plain an intention to place *Selection 44* in the context of modern dance music and that perhaps I may have just as effectively have labelled the piece 're-ordered by', 'paraphrased by', or 'collaged by', each of which would have adequately described the use of Haydn's material and each of which would have implied an association with an existing art form and that the association would help to validate and authenticate the work within a certain paradigm. Though the work clearly owes much to the minimalism of Reich, Glass, and others, one particular conscious influence was The Chemical Brothers, of which the opening of 'Come With Us' (2002) uses a repeating string sample that rises and builds tension towards a point of release. The compositional techniques share ground with minimalism but also, as in the previous example, with the *Sturm und Drang* techniques of the Haydn Symphony. For *Selection 44* a recording of the Presto was sampled on a computer bar by bar and remixed, much in the same way that a sample might be used in a dance track, before being transcribed back onto a score for live performances. So that, although no other music was added, neither was it electronic or to be performed in a club etc., it still contained the qualities of a remix. If there was any clever technique to it, then it lay in how the bars were re-ordered, so that they still made 'sense' harmonically and compositionally speaking and with regards to the parts for the performers so that their parts flowed in the way in which classical music usually does.

Defining Remix and Sampling

It is fair to say then that the work was 'sampled' and that it was 'remixed'. Tim Prochack in *How to Remix* (2001), though noting a multiplicity of viewpoints on remixing (and recognising a tradition that extends at least from Bach to Fatboy Slim) settles on the following property of a remix:

> Put simply, with no reading between the lines, it's a "version" that is an interpretation of another song but is still markedly different, and that's how remixing is different from the rest of dance music. Although dance samples other material left, right and everywhere, it usually does its best to conceal its source. A remix, by default, pays homage to its original source.
>
> (Prochak 2001, p.18)

So, what then relates *Selection 44* to a longstanding classical tradition? I am not proposing to tackle all sampling related issues in depth, such as the ontological issues regarding the identity of a piece (when is Haydn's *Symphony 44* no longer Haydn's *Symphony 44*?), or ethical issues regarding plagiarism or copyright law (the rights or wrongs of downloading DJ Danger Mouse's *Grey Album*, for example), or to make value judgements based on style (hence Palestrina and the Sugababes appear in the same category without need for apology). There is clearly a differentiation to be made between the amount of material that is used in employing a short quote through to using all of an existing work, so can these forms of music be categorised and placed into a meaningful typology? The construction of a taxonomy of types faces problems in an age where reducing entities to an essence is almost a criminal act, but we might hold that so long as we maintain scepticism over the borders of types, then categorisations have a useful function in commenting on the working of the practice under scrutiny.

Sampling and remixing in the modern era

Sampling, as a term, belongs to the modern era as it is directly related to the use of technology and there are key artists and works that, if not the first of their kind, are considered seminal in their influence on future work. The late '60s, early '70s Jamaica saw the rise of the 'Dub' mix, where producers deconstructed tracks often to reproduce pieces that concentrated on the bass and to add reverb designed to impress on the outdoor sound systems that sprung up across the island. Artists such as Lee Scratch Perry and King Tubby created seminal tracks, for example, King Tubby's Rocksteady Dub mixes Augustus Pablo's cover of 'Barbwire' with added samples from Nora Dean.

At a similar time, Steve Reich's early works, *It's Gonna Rain* (1966) and *Come Out* (1967) have a purity in their minimal concentration on material and are, in part, separated from dance by the lack of drums and electronic instrumentation, though not surprisingly dance artists such as Coldcut, DJ Spooky and DJ Freq Nasty, consider him a key influence. Likewise, Pierre Schaefer's *Chemin de fer* (1948) using samples of train sounds collaged into a soundscape, clearly uses the same process of taking recorded sound but the term 'sampling' was not widely employed by the Musique concrète artists of the '40s and '50s. There is a similarity between the use of concrete sounds as 'found sound' and extracts of existing music used as 'found sound'. In 2004, the BBC Proms commissioned Swedish composer Anders Hillborg's work *Exquisite Corps*, a Surrealist-inspired musical collage that used works of twentieth-century composers, as Hillborg describes:

> I tried consciously to combine disparate material from my own pieces as well as those of other composers. For example, there is a chord from Petrushka, a style quotation from Ligeti, a (hidden) quotation from Sibelius.
>
> (Hillborg 2004)

The collage technique extends back at least to the early twentieth century such as in Ives' Putnam's Camp from *Three Places in New England*, dating from 1914, that mixes well-known tunes and is inspired by the sound of multiple marching bands marching simultaneously. Musicologist J. Peter Burkholder, amazingly identified *14* types of uses of existing music just in the work of Charles Ives alone (see Burkholder 1985; and 1994). Bernd Alois Zimmerman became known for using collage techniques, the opera *Die Soldaten* (1960) being a key example; John Cage too, through chance processes collaged existing work such as in *Roaratorio* (1979). These works are still considered to be 'by' the composer, as the composer has acknowledged the authorship of the existing material used and the listener is aware of the acknowledgement. So it is with *Selection 44* that in my claim to be the remixer rather than the composer (as is the fashion on remixed tracks), there is the recognition that I have employed at least one compositional technique and have made artistic decisions in the process, but that not enough of the artistic decisions are mine (consider all the melody, harmony, orchestration etc. that is Haydn's) in order to be called 'the composer', though I must still consider myself as 'an' author here.

Retrospectively, I am aware of other pieces that gave me 'permission' to remix Haydn, in the same way perhaps that Cage considered Rauchenberg's white canvases as giving him permission to 'compose' *4'33"*. Nic Collins at the LIMTEC Conference in 2003 (LIMTEC) played his work *Still Lives* that uses a CD of a piece by Giuseppe Guami (*b*1540, *d*1611); by removing the mute function from the CD player and pressing and releasing the pause button, the CD looped producing instant minimalism from an existing piece. Less left to chance but equally minimalist is Michael Nyman's use of Purcell in *The Draughtsman's Contract* (1982) where short phrases are rearranged for more modern instrumentation. So far, I have chosen examples that by and large, (with the exception of some instrumentation), do not add new material composed by the 'composer'. But there are other categories that could begin to form the basis of a typology.

'Super' Compositions

In modern times there is a trend towards adding material to complete recordings; Kenny G adding his saxophone to Armstrong's cover recording of 'Wonderful World' is a noted example, as is Jan Garbarek's 'collaboration' with The Hilliard Ensemble combining saxophone with Renaissance and earlier polyphony in *Officium*. Other examples include Taverner writing new material over the top of Bach and David Fanshaw writing choir over recordings of African tribes (African Sanctus). 'Super' denoting 'above' or 'on top of', in this case.

Mash-up

The genre of Mash-up is recognised as mixing together two or more existing recordings. DJ Danger Mouse, in 2004, mixed The Beatles *White Album* with Jay-Z's *Black Album*, to produce the *Grey Album*. The White Album is used to provide all the drums and instrumental backing, whilst the Black Album uses an *a capella* version released by Jay-Z as the vocals and as the structure, so that the song titles on the Grey Album are as for the Black Album. The album is not available in all good record stores, and websites with downloadable tracks have been served 'cease and desist orders' by EMI/M Jackson. However, you could at the time buy the CD artwork with 'free' audio on eBay, but that was also put to a stop to with a 'cease and desist' order.

Medley

Medleys, such as the Elephant Medley from Baz Luhrmann's film *Moulin Rouge* coherently strings together U2 and Elton John but uses no, or very little, 'original' material. However, they can still be described as original in the same way Hiller's collage and *Selection 44* can be described as original, in that they are recognisable as the thing they say they are. Medleys, a mainstay of music hall and Gilbert and Sullivan overtures, tend to have a simple sequential structure.

Cantus Firmus

This is the use of a single line (such as a plain song) on which to base new work. For example: the Sugababes' use of Gary Numan's bass line from 'Are Friends Electric' (1979) in 'Freak Like Me' (2002) shares a compositional strategy with 'early' music such as in thirteenth-century motets.

Chord Progressions

These are works that take existing and familiar chord progressions, such as All Saints' 'Never Ever' (1997) that employs the 'Amazing Grace' progression. Or the I, IV, V progression of 12-bar blues as the established basis and defining factor for innumerable compositions.

Cover Version

A cover version is a new version of a song previously recorded by another artist. Christina Aguilera and Missy Elliot's 2004 remake of Rolls Royce's 'Car Wash' (1977) has made artistic decisions in the alteration of the original, and the distinction between interpretation, composition, production, and arrangement become blurred.

Pastiche

Pastiche or parody of a style may not directly use exact copies of intervals, rhythms etc., but there must be elements that are similar in order for it to be pastiche. For example, a modern Hollywood blockbuster will draw on known signifiers and allude to known compositions. In *Troy* (2004) composer James Horner calls on the march rhythms and horn calls that signify armies and triumphal marches using a language that could be said to be 'borrowed' from, or at very least influenced by, the Sanctus in Britten's *War Requiem* (1961).

Quotation

This seems to be the least invasive use of existing material, where a small amount of material is quoted. Modern examples might be Tippett's quote of Beethoven 9th Symphony in his *3rd Symphony*. An earlier example would be Berg's quote in the Lyric Suite (1925) Of Zemlinsky's 'Du bist mein augen' from his *Lyric Symphony* (1923). A recent example is Eminem's use of a 'lullaby' ('hush little baby') in the 2004 album *Encore*.

Chronology

So far, I have concentrated mainly on modern and twentieth-century uses of existing material, whereas the technique of basing a new composition on existing material was the norm, rather than the exception, in earlier forms of Western music. In early music, where there is little or no notation to exemplify contemporary practice, there are difficulties in discussing authorship, though it is not fantasy to speculate that there must have been songwriters and tunesmiths who wrote original material as troubadours such as Adam de la Halle in the thirteenth century would have done.

Histories of Western music pay attention to plainsong and its use as the basis for adding newly composed parts. Early records of organum, such as exemplified in the florid style of the St Martial School (early twelfth century) show how the plainsong melody was slowed down ('time stretched') in the lower voice, the Vox Principalis, with an original and florid part, and the Vox Organalis written over the top. In the Notre Dame School of the late twelfth century Leonin, in *organum per se* and *discantus* style, the tenor elongates the plainsong and adds new material on top. In Western classical tradition this was normal practice. Existing lines were added to existing lines on top of which new lines (duplum, triplum etc.) could be added, so that one composition was written on top of another, which in turn, becomes the basis for layering another composition on top. For example, Adam

De La Halles' 'Jeu Robin M'aime' is built on a plainsong 'Portatre' and in turn a triplum Mot ne fu griefs was subsequently added on top. Organums, motets and later some cantus firmus masses, use the plainsong as a compositional washing line for the pegging out of new material on top. In essence then, this is perhaps no different to the Sugababes using Numan's bass line as a cantus firmus, and the compositions, though many hundreds of years apart, share a common, identifiable, compositional technique.

Renaissance

In the early Renaissance the cantus firmus was based on extent plainsong and is still in use, but increasingly musical lines are written simultaneously, so that all music is original and the concept of the composer as the author becomes more firmly established. Clearly this is a 'modern' technique for composing. The sixteenth-century theorist Pietro Aron notes how 'modern' composers were beginning to eschew the practice of writing one part on top of another in favour of writing parts simultaneously:

> If you consider only part by part, that is, when you write the tenor and you take care only to make this tenor consonant, and similarly the contrabass, the consonance of every other part will suffer. Therefore the modern have considered this matter better, as is evident in their compositions for four, five, six and more parts. Every one of the parts occupies a comfortable, easy, and acceptable place, because composers consider them all together.
>
> (Pietro Anon Venice 1524, in Grout 1988, p.209)

With this 'new' technique the need for the existing material as a washing line diminishes and the dominance of the sole author as the great composer emerges. Although now, in the twenty-first century, this idea is being eroded and 'birth' has been given to the reader at the expense of the author (Barthes), but we might still expect, for example, proms commissions to be by *a* composer. We are not yet in a position where it would be normal practice to accept a piece as co-authored, for example, by Taverner and Turnage as we have come to recognise Lennon and McCartney as accepted joint geniuses. Modern art has already accepted joint authorship (Jake and Dinos Chapman, Gilbert & George), but the traditional mediums of classical work, orchestras, opera etc. are rarely joint collaborations in compositional terms. Though this is beginning to emerge; in the '90s Sandy Goehr used an unfinished Monteverdi opera *Ariadne* and completed it in a fused Goehr/Monteverdi style. And the ENO worked with the ADF (Asian Dub Foundation) on a new opera. Leeds Fuse Festival saw that one work in Fuse04 (2004), under the artistic direction of Django Bates, had 60 authors.

The Baroque, Classical, and Romantic periods show less reliance on the use of existing material as genius author dominated, but the practice was still accepted and commonplace. Bach harmonised the Lutheran hymns for functional music. Tunes were drawn from a common pool, such as that shared by Handel's 'And With his Stripes' from *The Messiah* and the Kyrie from Mozart's *Requiem*, each of which uses the material in different ways. The Variation on a Theme becomes a common form. Beethoven wrote variations on the British National Anthem for example, where the use of existing material falls into the category of being 'based on' something but for the purposes of creative developments rather than as an unchanged foundation. In Romantic times the art of paraphrasing returns in a different context to the Medieval use of paraphrase mass, whereby compositions were arranged or abridged for piano such as the Liszt arrangements of Mozart's *Requiem*, opera arias, Beethoven Symphonies, etc. Rather than remixes these might best be described (in modern terms) as cover versions. Some addition to existing work can be found. For example, Gounod's *Ave Maria* (1891) built on top of Bach's *1st Prelude in C*. This follows the model of the motet and fits into the category of 'based on' along with the Sugababes and cantus firmus mass.

Might it be revealing to put the types in order of the amount of existing material used and create a taxonomy of the types? Would a table that considers works on an individual basis show anything more informative about the use of existing material? For this individual works will need to be compared to each other; for example, *Selection 44*, which uses only material from a movement of a Haydn symphony, has more existing material than Berg's *Lyric Suite*, which includes a short quote from Zemlinsky. Gounod's Ave Maria is about 50 per cent Bach, 50 per cent Gounod, depending on how the observer chooses to weight the contributions, so that it uses more than the *Lyric Suite* but less than *Selection 44*. The amount of existing material used in individual pieces could be a quantitative measurement. However, this does not go any way towards measuring levels of originality or creativity, the search for which remains just as unquantifiable and renders such comparisons fairly meaningless. So, although this study began as a search for a history of remixing and to trace the lineage of *Selection 44*, what emerges, and perhaps what is most valuable from this study is the identification of compositional methods using existing material that exists through all ages. So that, regardless of value judgements, the Sugababes use of Gary Numan's bass line shares a technique with a cantus firmus mass from the late twelfth-century works of Leonin and Perotin, and the distance between twelfth and twenty-first centuries is reduced to nil.

Part Four

Type 4. Deproduction

The Collective Disappearance of Production

17 On Writing Every Song

Robert Wilsmore and Phillip Brady

We have proposed the possibility that music production is finite rather than infinite. Even if the latter is the case, we can confine ourselves to a generalisation that places limits on how we engage with music. Tunes can be long, possibly infinitely long, but none of us can listen to the whole of that composition, if the universe ends, the song ends. 'The Song of a Thousand Songs' is, sadly or fortunately, never going to be heard by any one individual in its entirety (if it has an entirety). But to concern ourselves with our everyday humming of a tune, particularly one that would meet the criteria of Winnie the Pooh to whom a good tune is one "such as might be hummed successfully to others", we can, for exploratory purposes, reduce the potentially infinite down to a short ditty, something manageable so that we can encompass all short other ditties before moving on to larger ones, and then to musical things beyond sequences of pitches. So, as we have already focused on a particular pithy phrase in Toast theory in a previous chapter on removing non-sonic signifiers from the semiotics of endings, we will return to the opening phrase of Beethoven's 5th Symphony as the sequence upon which we will draw our parameters for our attempt at producing 'everything'. The tune is instantly recognisable, what's more it is short, and it can be placed within the parameters of the tonal and rhythmic infrastructures from the symphony from which it is drawn. Evidently this is not 'everything', but it is everything within the defined parameters, and hence a start in moving towards the wider everything.

We set our task on working out how to produce all possible variations of sequences within the parameters derived from the opening refrain from Beethoven's 5th Symphony, the precise details of which we will delineate in the following chapter. One of these mathematically generated variations, set in a logical order from first to last, will be the tune that Beethoven 'wrote', that is, the opening two-bar melody appears at a certain point within the millions of permutations and this point is directed by how we proceeded with regards to the logic of the sequence. We put Beethoven 'wrote' it in quotations to emphasise our position that if all such variations are already possible, rather than *creating* them the composer *discovers* them and then places their flag on that territory. And this takes us back to the central tenet of this book. It may be that one explorer (composer) has the dominant claim to discovering, and therefore claiming ownership of, the territory but on the whole lots of explorer–composers seem to be trying to make their flag fly highest and strongest in the same territories or in overlapping territories that have no defined border except for those that have been drawn in the sand by those with the flags. Returning to our immanent project, our task is to 'discover' all the variations within our parameters (within our lines drawn in the sand) and to locate Beethoven's 'da da dah' tune within them. Consequentially, we will discover all other tunes within those parameters, some of which will already have flags on them. How we will deal with that 'Monopoly board' of minding which territory one lands on we are not sure but will speculate on the ideas. Similarly, the temptation to plant a flag on any territory so far unclaimed is tempting but that is not our intent. A life spent battling for ownership in the

DOI: 10.4324/9781351111959-21

law courts does not sound very inviting. We will benignly 'categorise' the tunes, and, as we are working with computer programmer Neil Ward to develop an interactive application for this, Beethoven's tune will be labelled on the BWW (Brady-Ward-Wilsmore) system, like the Bach-Werke-Verzeichnis BWV system of categorising Bach's works, or the manner in which stars, too numerous to name, end up with impersonal numbers on the Hipparcos star catalogue where Betelgeuse (Alpha Orionis) is merely HIP27989.

But numbers are also human things, and we must not forget that mathematicians are human too. There should be nothing worrying in our mathematical approach to 'composing' all tunes. We are not dehumanising but merely demystifying production. If the 'magic' is lost in the process then let it be replaced with 'wonder' instead, the wonder of the incredible beauty of mathematics. Mathematicians often refer to music, the links between ratios and intervals is clear, the mathematical beauty of the harmonic series that is inextricably linked to the home comfort of the major chord that its first overtones creates. We chime with the universe most probably because we are made of the same stuff and structured by the same laws. In his discussion on infinity in the book *From Here to Eternity*, mathematician Ian Stewart writes that:

> A musician would be horrified if his art were to be summed up as 'a lot of tadpoles drawn on a row of sheet music' but that's all the untrained eye can see in a page of sheet music. [...] Mathematics is not about symbols and calculations. These are just the tools of the trade – quavers and crotchets and five-finger exercises. Mathematics is about *ideas.*
>
> (Stewart 1987, pp.1–2)

So, here is an idea, a mathematical, musical, and philosophical idea, that is that we will set about the task of writing every possible tune. Within limits, at least (so *not* 'every' tune, of course; not yet at least). It will be one thing to calculate the number of variations possible, this requires a clear methodology, but such calculations are not as challenging as they might seem. But it is another thing to actually *produce* the sequences as notes ('tadpole' representation or otherwise) in a manner that means that they can become the sound, which is, after all, the music rather than a representation of the music. This is likely to require translation into computer programming that can then generate the sounds (and the score) if we are to get anywhere near being able to produce the millions of variations that we find. As we have just said, some calculations are relatively straightforward, for instance calculating the number of possible tone rows. We can ignore duration here and keep the parameters simple with regard to each of the 12 notes of the chromatic scale appearing only once in a sequence of 12. The simple calculation of factorial 12 (12!) produces 479,001,600 rows, although it does not take into consideration how we might view replicated intervallic sequences as being the 'same' (i.e. not unique). Starting the sequence on a different semitone counts as a different variation in this approach, and the starting points are 12 according to the abstract notion of the name of the note, which is assumed to be that of an actual pitch in terms of frequency in Hertz according to our general fixed tempered scale with A=440.

Our first task is quite specific with regard to its restrictions, i.e. defining the parameters of Beethoven's theme, but there are other attempts towards the same agenda. Rhett Allain's writing, in his article 'How Many Songs Can There Be' in *Wired* (2015) demonstrates a mathematical approach to writing all three note sequences starting from the viewpoint of the notes coming from a seven-note scale (rather than the chromatic 12) using the equation $N=7^3=343$. His starting point in the world of songs comes from a consideration of whether Sam Smith's 'Stay With Me' is based on Tom Petty's 'I Won't Back Down', and this leads him to make the calculations with regard to generating all possible three-note sequences. In the correspondence on the website following the article, a reader comments the following to which Allain then responds:

> [Reader] There is a huge flaw in your plan. If you write a song that reproduces all the other note combinations, won't you also reproduce the three notes from He's So Fine and also be sued by Bright Tunes? Won't you open yourself up to all the previous songs that have been written suing your pants off?
>
> [RA] That is a great point. Thank you for contributing to the seriousness of this discussion [...] Yes, in my plan I will likely get sued. However, if I sue more than I get sued I should be fine.
>
> (Allain 2015)

So, if we are to write every song then we will also have to contend with other musicians and mathematicians who claim to have ownership of every song. Once more, like our previous example of Lana Del Ray, Radiohead, and The Hollies, we are back to the unsolvable problem of origination. Nearer to the task of our own search is one that Allain points us to writing "But of course I am not the first person to calculate the number of different songs. Rachel Hall (a music professor) used 16 beats (7 notes each) to get a value over 30 trillion" (Allain 2015). Again, our project is not just the calculation of the number of permutations but the *actual manifestation* of all of them, and that project changes the mathematical approach in ways that are not anticipated when calculating numbers of permutations only. There are 'unknown unknowns' that need to be tackled in the making of the tunes. Collectively as one world, as contributors to 'The Song of a Thousand Songs' we are having a fair go at the task of 'everything' but we are far short of arriving at that finite point that completes the state of deproduction. And we have to say that, with our small range that produces millions of tunes, then our immediate thoughts turn to whether this proves the worth of composers, whom with the odds firmly stacked against them with regard to picking a winning tune at random, they pull out 'a tune in a million'. This position switches the composer back into the realms of myth and magic, back to the label of 'genius'. Maybe a composer could get lucky, perhaps they could be the lottery winner who randomly picks a tune that's a big hit. We might even attribute this to the 'one hit wonders' of our industry, but the Mozarts and McCartneys of the world seem to be able to catch us with their tunes time and time again, and statistically no one is likely to be that lucky.

Let us speculate for a while. Can we envisage a time then when our record production industry could be so automated that we could indeed turn it into a lottery? In this mode the songs would be generated systematically, and the lottery system only made available those songs that are not already 'owned'. Players could select which of the songs they wish to 'purchase' as their lottery ticket or they could choose a random 'lucky dip' option. The songs are then put out to market and the owner of the number one song ticket gets the revenue. Or if we take ownership out then it might be more like a horse race where we bet on the winning song, and of course, we already do this and, in the UK at least, betting on the Christmas number one is something of a tradition. In October 2017 Ladbrokes had Wham's 'Last Christmas' as 4/6 favourite, with Ed Sheeran on 3/1 odds and 'the eventual winner of The X Factor' at 6/1. This last bet is of particular interest to this study because it is already contingent, that is, one places one's bet without knowing what the song will be (much like our hypothetical lottery above) and one will know the potential winners of the show but not the *actual* winner. But statistically from 2005 to 2014 X Factor artists accounted for seven out of the ten UK Christmas No.1s, the other three comprising of two charity songs and one 'anti X Factor' song, Rage against the Machine's 'Killing in The Name', driven there by a social media campaign to draw some (supposed) authenticity back to music and defeat the X Factor hit machine (by beating X Factor's Joe McElderry to the top spot). The 'Rage' entry is clearly a backlash against the automation that we are considering here but again we do not consider The X Factor or our systematic mass production approach to be an attempt to undermine our humanity but to be part of our authenticity, in much the same way that Allan Moore describes second person authenticity as the 'authenticity of experience' (Moore 2002). If we can overcome the luddite fear that automation destroys music then the 'wonder' of music can return. In the end it was the 3/1 odds Ed Sheeran that made the UK number one in December 2017. Caroline Davies in the *Guardian* online writes:

> The 14th series of The X Factor kicked off with the lowest viewing figures since the show first began in 2004, although it was still the most watched programme of Saturday night. The return of Simon Cowell's ITV talent contest attracted an average of 6 million viewers (a 32% share), peaking with 6.9 million (36%). This compares to '10.8 million in its heyday'.
>
> (Davies 2017)

As such, its odds for the Christmas number one were likely to reduce and the show had even changed its release timing of the winning single, possibly in recognition of this. In the end the 2017 Xmas slot went to Ed Sheeran's 'Perfect', which topped the UK charts for three weeks. So, originality won in the end. Or maybe not; in January of 2018 avclub.com reported "It's a new year, and with that new year we now have another new lawsuit in which someone has accused Ed Sheeran of ripping off their song" (Barsanti 2018) with the article heading "Ed Sheeran is getting sued again". Although not about the song 'Perfect', it once more brings into question the idea that an artist might be any more original than a machine. The originality debate seems to continue on separate pathways, in the philosophical tradition the Platonic form disappears through Baudrillardian Simulacra only for the concretion of objects to be reinstated in Graham Harman's Object Oriented Ontology (2019).

From our standpoint the frequency of media reports of lawsuits with regard to authorship is not an increase in plagiarism but evidence that deproduction is beginning to speed up. Computer programmes have also claimed to *reduce* the formulaic and help the lesser known artists gain credibility. Music Intelligence Solutions inc. produced their Uplaya.com site in 2009 not so that hit records could be manufactured but as CEO David Meredith is reported as saying in Marketwired:

> "As major record labels pull back on promotion for new artists, uPlaya provides a needed service by shining a spotlight on the songs of talented musicians from around the world who benefit from an objective validation of their work," said David Meredith, CEO of MIS/ uPlaya. "A strong Hit Potential Score coupled with uPlaya's viral marketing tools can help level the playing field for the millions of artists working hard to break out," added Meredith.
>
> (Marketwired.com 2009)

The Music Genome Project run by Pandora Media Inc. in the US, made explicit statements about the focus on the human and the individual that their technology serves:

> Pandora is the world's most powerful music discovery platform – a place where artists find their fans and listeners find music they love. We are driven by a single purpose:

> unleashing the infinite power of music by connecting artists and fans, whether through earbuds, car speakers, live on stage or anywhere fans want to experience it.
>
> (Pandora N.D.)

Of course, this is the company website and there are market forces at play in the human spin with which the narrative plays. But its point is a fair claim for many technologies, that is, it enables us to be more engaged with the activities that we want to be engaged with as humans.

So, as our proposed computer generation of all tunes will produce too many for actual people to listen to and choose from, it will need further software to select the ones that we will want to hear. Perhaps we could apply the smart and fun 'hit equation' developed at scoreahit.com to our tunes, or we could skip straight to joining in the race to making an AI computer programme that writes 'good' tunes in the first place (leaving out the millions of not very good ones), in the manner in which Sony CSL research lab's 'FlowMachines' might be doing as we enter the 2020s. Indeed, it is possible that the 2020s might be the age where these 'machines' are as accepted as any other songwriter. Give it a few more years and AI should be able to write music on the spot based on biometric feedback that we *will* like. Of course we will, the AI programme has analysed our data and knows how to reproduce the '*like*' affect in us by providing the music that produces a favourable response. It will be music that we will like. Guaranteed or your money back. But that is something only a report from the future could tell us.

However, to come back the task in hand, our purpose is not to write hit songs, it is not to place value on them, it is to write (or rather to 'categorise') *every* song, regardless of whether anyone or any computer software thinks they are good or not. Pedantic and prosaic that maybe, but that is our task. Write down every tune within limited parameters based on the opening two bar melody of Beethoven's 5th Symphony. We do not consider ourselves to be composing the tunes; we are categorising and ordering them in a particular way so that they can be set out in a logical sequence. When we arrive at Beethoven's 5th tune, it will simply be the next tune in the sequence, and the sequence will then continue on past it. But we will put a flag on it saying, "Beethoven was here!", given that is what we set out to do.

18 The Mathematics of Writing Every Tune

Phillip Brady and Robert Wilsmore

Method for Calculations

The tricky part to actually writing 'every tune' is not calculating how many possible tunes there are but in finding a sensible ordering system that will allow us to produce them. Calculating the number of possible permutations can be straightforward when given precise parameters but our task involves the *application* of the mathematics in such a way as to produce a logical system that can be translated into the tunes themselves; or at least as notes on staves that can then be sounded, the point being that they will be understood as tunes in the manner in which we would normally do so. Hence there is, in one way, no difference between the numbers in the boxes on the vast spreadsheets (that we ultimately produced through applying the maths), to the notes on a musical stave; they are both merely representations of the sound as opposed to being the sounds themselves. It could be said that we are just replacing one notation system with another and in one sense this is true, but the mathematical version we present places emphasis on categorisation, position (a location address), and logical progression, with the order of progression fixed by us in this system from start to finish (perhaps that makes it a composition after all?). As we wish to engage in the everyday world from an everyday perspective, we have at times, for exemplification here, translated our numbers into more familiar musical notation: notes on staves. Perhaps ironically, in order for these 'notes on staves' to sound as tunes, they will need to go through a process of conversion to binary so that they can be played from a computer. Unless of course we can persuade 'live' musicians to play them. Were they to perform the sequence from start to finish it would last about five years depending on what tempo is set. Our first task is to set out the aim, the parameters, and the working methodology.

Aim: Our aim is to produce, in a systematic manner, every possible tune that can be made within limited parameters derived from the opening two-bar phrase of Beethoven's 5th Symphony, and then to locate Beethoven's melody within that system. Our aim is not merely to count the number of possible tunes within the parameters but to produce a system that can place them in a logical sequence, and that each tune can be given a label within that system. It is planned, following on from this publication, that we will work with computer programmer Neil Ward in order to create an application capable of presenting this system on computer, whereby we can locate and view each tune on the Brady-Ward-Wilsmore (BWW) system, with each tune having its own address or set of coordinates.

Parameters: We have outlined the approaches to pitch, key signature, duration, and rests and suggest some of the limitations of the approach.

Pitch: We have kept to one octave of eight notes, which ranges from one tonic to the tonic an octave above. In the case of C minor (the key of Beethoven's 5th Symphony), this is from C to C an octave above, inclusive of both. It holds that the *numerical*

DOI: 10.4324/9781351111959-22

possibilities are the same for all eight note sequences, hence it can be abstracted out numerically. We refer to these sequences as 'tone rows' noting that we are not referring to the 12-note tone rows of twentieth century serial music.

Key signature: We proceeded initially as if the key is C Major, that is, ordering (with the complexity of the rests thrown in) the natural notes C, D, E, F, G, A, B, C. With regards to the final 'tunes' that are created, the necessary flats are then placed as a key signature (B^b, E^b, A^b in the case of C minor) which affects the resultant tunes (as it affects the order of the intervallic relationships). But for our purposes we proceed without the need for a key signature in the first instance and operate as for generating all tunes within C Major as the 'retrofitting' of C minor does not affect the rigour of the systematic approach. The location of Beethoven's 5th (at least in its original position as opposed to transposed versions) will appear in the same location as for the version in C Major that goes G, G, G, E (as opposed to G, G, G ,E^b), the exception being that the first will be referenced in the C Major format of BWW-C(n) and the latter minor version (with the C key signature) as BWW-c(n), with the capitalised 'C' representing Major and the lowercase 'c' as minor as is normal in classical music theory.

Duration: The shortest duration we have allowed is a quaver (⅛ note), the longest is a semibreve or whole-note (effectively two minims tied across two bars of 2/4). The whole phrase lasting two bars of 2/4, that being 4/4 or '1' in total. We made the decision to limit the durations between quaver and whole-note to simple additions of the ⅛ note, hence the only permissible notes are: ⅛, ¼, ⅜, ½, ⅝, ¾, ⅞, and 1. Or for the musically minded: quaver, crotchet, dotted crotchet, minim, minim tied to a quaver, dotted minim, minim tied to dotted crotchet, and semibreve. The decision is logical but clearly excludes many common durations such as semiquavers or triplets, each of which might increase the possible permutations by millions or billions.

Rests: Given that the opening of Beethoven's 5th Symphony includes rests (the very first 'note' of the symphony is a quaver rest), we choose to include rests to the same value of the durations. The inclusion of rests was to prove a significant factor in how the system was developed, that is, that rests are considered as part of the rhythmic system as opposed to the pitch system. This is one example of where the application of mathematics required a solution that is not necessary in the calculation of 'how many' tunes there are.

Limitations: This is obviously very restricted in its remit. Most tunes are longer than this, and the resultant tunes in our category will exclude anything that falls outside of these parameters. For example, anything that has a triplet, or a semiquaver, or an accidental outside of the key, will not be categorised here. The obvious conclusion being that we are very far away from writing every tune, but we are writing every tune possible within these parameters and that is our task.

Calculations: For ease of calculation we consider the note lengths to be all integer values; 1 representing the shortest length of note (⅛) to 8 being the longest (the semibreve). As well as avoiding the need for working with fractions, this has the advantage of simplifying the extension to longer tunes.

Rhythms: On the face of it, this seems like the most straightforward part of our calculation. All we need to do is find all ways to split up 8 into sums of integers (partitions of 8) and then consider all permutations of these partitions. There are 22 partitions of 8 and with permutations the count is 128. By thinking of each element of the sum as a note length within the 8, our permutations cover all rhythms with no rests. For example, the partition [1+1+1+1+1+1+1+1] represents the single rhythm with 8 quavers, whereas the partition [2+1+1+1+1+1+1] represents rhythms with a crotchet and 6 quavers. However, this partition has 7 permutations, so in effect represents 7 distinct rhythms.The inclusion of rests brings some difficulty. One possible solution would be to assign a zero pitch to a duration to indicate no sound. The problem with this approach is that we would have several tunes in our list that were essentially the same. For example, 2 quaver rests and 1 crotchet rest would 'sound' the same. Our solution is to restrict all rests to having a duration of 1, thus in the example above only 2 quaver rests would be allowable.

In order to understand how the rhythms are formed and how the system may be extended we will start by considering the very simplest of rhythms.

Layer 0: Rhythms of length 0.

This empty, or null, set with no notes and no rests is a strange but important foundation on which to build all rhythms of any given length.

Now we extend the rhythm by one quaver.

Layer 1: Rhythms of length 1.

- First, we add on all rhythms of length 1 containing no rests. There is only one of these – a single quaver.
- Next, add a rest to all the rhythms built so far. Since we only have the null set, this only introduces one rhythm – a single rest.

Thus, there are two possible rhythms of length 1: a single quaver (1); a single rest (0).

Our building continues to its second layer in a similar manner.

Layer 2: Rhythms of length 2.

- All possible rhythms of length 2 not containing rests added [there are 2 of these – (11) (two quavers) and (2) (one crotchet)].
- A rest is added to each of the rhythms from layer 1 [(**0**0), (**0**1), (1**0**)].

The total is now 5, not 4, since the rest added to the single quaver from before can appear either before or after it.

Layer 3: Rhythms of length 3.

- All possible rhythms of length 3 not containing rests added [there are 4 of these – (111), (12), (21) (3)].
- A rest is added to each of the rhythms from layer 1 [(**0**00), (**0**01), (01**0**), (10**0**), (**0**11), (1**0**1), (11**0**), (**0**2), (2**0**)].

We can continue in the same way until the rhythm is of the desired length.

Layer (r): Rhythms of length (r).

- Add all rhythms of length (r) with no rests.
- Add a single rest to each of the rhythms from Layer r-1 (in all possible positions).

At this point there are two considerations to be made:

1. How can the new permutations of length (r) be produced in an ordered fashion?
2. How should the new rest be placed in all possible positions?

Partitions and Permutations

We now search for a way to produce all permutations of partitions of an integer (length) r given those for integer r-1. For this it is useful to divide the partitions according to the number of elements (notes) used in ascending order. For ease of notation we will omit the + signs when listing partitions of an integer, for instance, 112 will be used to represent the partition 1+1+2. Since we are interested in permutations, the partition 121 will be considered to be different from 112.

Partitions of integers (lengths) from 1 to 5 are summarised in Table 18.1 for easy reference, followed by Table 18.2 counting the number of possibilities.

Table 18.2 appears to show Pascal's triangle, a pattern familiar to many. Values in any given cell can be found by adding the values from the cells immediately to the left and diagonally left and up (except for the top row, which is always 1). Why should this be so? New permutations are formed by adding a 1 to the start of all those from the previous partition with length 1 less (e.g. (12) → (112), (21) → (121) in Table 18.1 above) and copying those from the cell to the left with the first digit increased by 1 (e.g. (111) → (211). Using this systematic approach, the partitions in each cell will automatically be in numerical order and the method allows us to extend to partitions of all integers including their permutations.

Table 18.1 Partitions of integers

Partitions		*Integer (length of sound)*				
		1	*2*	*3*	*4*	*5*
Number of elements (notes)	1	1	2	3	4	5
	2		11	12	13	14
				21	22	23
					31	32
						41
	3			111	112	113
					121	122
					211	131
						212
						221
						311
	4				1111	1112
						1121
						1211
						2111
	5					11111

Table 18.2 Count of partitions

Count of partitions		*Integer*				
		1	*2*	*3*	*4*	*5*
Number of elements	1	1	1	1	1	1
	2		1	2	3	4
	3			1	3	6
	4				1	4
	5					1

It is well known that the values in Pascal's triangle can be calculated using the expression

$$\frac{n!}{r!(n-r)!} = \binom{n}{r}$$

where n is the column number and r is the row number (with the convention that row and column numbers start at 0).

To make use of this for our situation we introduce the following variables:

n: the total length of the rhythm;
l: the total length of notes sounding (i.e. the number partitioned before adding rests, labelled 'integer' in the table above);
s: the **number** of notes sounding (labelled 'Number of elements [notes]' in the table above. Note: not the length of notes sounding).
Note that the number of rests = $n - l$
From this we see that the number of rhythms (including permutations) of a length l with s notes is

$$\binom{l-1}{s-1} = \left(\frac{(l-1)!}{(l-s)!(s-1)!}\right)$$

Rests

Now that we are able to list all rhythms without rests of length 8 or shorter in a sensible order, we are ready to introduce rests. How many rests? All rests are considered to have length 1; rhythms with notes sounding of total length 1 ($l = 1$) will need 7 rests, length 2 will need 6 and so on. Notice that the number of rests is (8 – total length of notes sounding = 8 – l). The length of each rhythm will now be (total length of sound + the number of rests). To be systematic we will add zeros to each rhythm in turn, making sure that all possible positions for the zero are used. A sensible order would be to keep rhythms listed in numerical order starting with all rests at the beginning and finishing with all rests at the end.

To generalise we introduce an addition variable:

r: the number of rests

In general, if n represents the total length of the rhythm but the rhythm only has s notes sounding, of total length l, in order for all listed rhythms to have length n we must add $r = (n - l)$ rests to each restless rhythm making the rhythm $s + r = s + n - l$ elements long.

The number of ways to position $(n - l)$ items in a set of $(s + n - l)$ characters is

$$\binom{s+r}{r} = \binom{s+n-l}{n-l}$$

Now we list in a table all the possible rhythms of length n with cells (s,l). Each cell in the table will contain $\binom{s+n-l}{n-l} \times \binom{l-1}{s-1}$ rhythms.

For example, using the cell containing (112), (121), (211) in Table 18.1 above: The table is the basis for rhythms of length 5 so $n = 5$. This cell has 3 sounding notes of total length 4, so $s = 3$ and $l = 4$. Now $r = n - l = 5 - 4 = 1$. Hence, we must place 1 rest in a rhythm of total length 4 ($= s + r$). There are four ways to do this for each restless rhythm leading to [(0112), (1012), (1102), (1120); (0121), (1021), (1201), (1210); (0211), (2011), (2101), (2110)]. These should then be put into numerical order (i.e. (0112), (0121), (0211), (1012), (1021), (1102), (1120), (1201), (1210), (2011), (2101), (2110)).

If we add these together down the columns and along the rows, we find that the number of rhythms (with 1 added to account for the rhythm with only rests) is

$$1+\sum_{s=1}^{n}\sum_{l=1}^{s}\binom{s+n-l}{n-l}\times\binom{l-1}{s-1}$$

For the parameters set out above, $n = 8$ and the number of rhythms is

$$1+\sum_{s=1}^{n}\sum_{l=1}^{s}\binom{s+8-l}{n-l}\times\binom{l-1}{s-1}=1597$$

By way of example, Table 18.3 shows how the rhythms may be listed for a tune of length 4 (quavers), i.e. one bar of 2/4.

Consider a rhythm with 2 notes sounding ($s = 2$) with total length of sound of 3 quavers ($l = 3$):

Number of permutations for sounding notes =

$$\binom{l-1}{s-1}=\binom{2}{1}=\frac{2!}{1!1!}=2 \text{ i.e.}(12)\text{and}(21)$$

Table 18.3 Rhythms for a tune of length 4

Length = 4 (n) *Total = 34 rhythms*		*Number of rests (r) / length of sound (l)*				
		4/0	*3/1*	*2/2*	*1/3*	*0/4*
Number of notes sounding (s)	0	0000				
	1		0001	002	03	4
			0010	020	30	
			0100	200		
			1000			
	2			0011	012	13
				0101	021	22
				0110	102	31
				1001	120	
				1010	201	
				1100	210	
	3				0111	112
					1011	121
					1101	211
					1110	
	4					1111

Number of ways of positioning rests =

$$\binom{s+n-l}{n-l} = \binom{3}{1} = \frac{3!}{1!2!} = 3$$

Note the 2 groups of 3 rhythms relating to 2 partitions each with 3 possible positions for the rest. These would then need to be put into numerical order (i.e. (012), (021), (102), (120), (201), (210)).

Pitch / Melody. Now that the rhythm has been fixed it is a simple matter to assign a pitch to each of the notes. If we assign to each pitch a value between 0 and 7 the melody can be indicated by a number written in base 8. The notes can be assigned to digits arbitrarily, but for ease of reference we suggest C = 1, D = 2, up to C' = 0>. This way the tonic is associated with 1 as we might expect. The number written in base 8 must have the same number of digits as the number of notes in the rhythm but since our rhythms have all been classified with regard to number of notes sounding (s), this is easy to achieve.

The number of tone rows associated with each cell in Table 18.3 is 8^s where s is the number of notes sounding. Therefore, the total number of one octave tunes of length 8 is

$$1 + \sum_{l=1}^{8} \sum_{s=1}^{l} \binom{s+8-l}{8-l} \times \binom{l-1}{s-1} \times 8^s = 83739041$$

Where Is Beethoven's 5th?

All 'tunes' with the parameters outlined above can be classified in a three-dimensional array (two dimensions [s and l] for the rhythm and a third for the melody). Each tune will be referenced in the Brady-Ward-Wilsmore system as follows:

BWW–[*n*,*s*,*l*](*p*)-{c *or* C}*number*

BWW – prefix to denote part of the Brady-Ward-Wilsmore system.

[n,s,l] – n represents the total number of quavers to fill (this is included to allow the system to be extended beyond our 8 quaver starting point); s the actual number of notes sounded; l the total duration (in quavers) of notes sounded.

(p) – the position in the cell of all rhythms with s notes sounding for l quavers. This is a number between 1 and $\binom{s+n-l}{n-l} = \binom{l-1}{s-1}$

{C *or* c} – upper case 'C' represents Major; lower case 'c' denoting 'minor' version.

number – a number in base 10 which, when converted to an s digit number in octal (base 8), would represent the pitch of each note by a single digit (0=C', 1=C, 2=D etc.).

For our purposes the opening two bars of Beethoven's 5th has overall length 8 (n = 8), 4 notes sounding (s = 4), and 7 quavers of sound (l = 7). It can therefore be found in row 4 and column 7 of the rhythm table; the rhythm 01114 is the first entry in this cell. So the rhythm ([n,s,l](p)) would be [8,4,7](1). The melody ('minor' version) is represented by the 4-digit octal number 5553, which when converted to base 10 is 2923 ($5\times8^3+5\times8^2+5\times8^1+3\times8^0$). So, Beethoven's tune in our Brady-Ward-Wilsmore system is:

BWW-[8,4,7](1)-c2923.

Its position relative to tunes surrounding it is shown in the diagram (Figure 18.1). Since the rhythms sit in an array, we should specify a playing order. The arrow indicates the

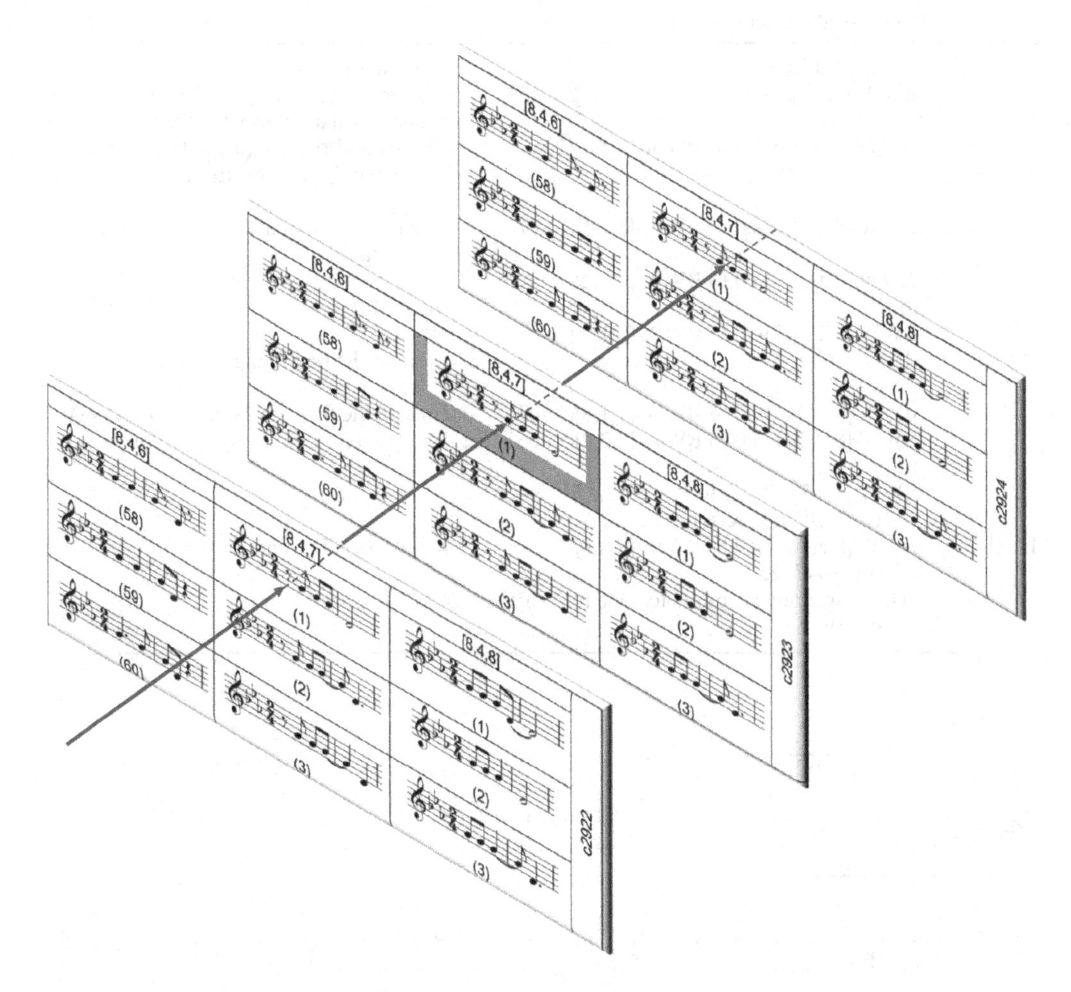

Figure 18.1 "Beethoven was here!"

progression route, for example for a linear 'performance', in this case showing progression through the [8,4,7] rhythms. All possible melodies for rhythm [8,4,7](1) are played from c0 to c4095 before moving on to all melodies for [8,4,7](2).

Performance Order

Here we outline more formally our order for 'performance'. After an initial two-bar rest (the 'null' tune), we propose starting with $s = 1$ and working across the columns increasing the value of l from 1 to n. All possible sequences of notes will be played in ascending order for each of the individual rhythms in any cell. This order will see early rhythms with rests and shorter notes at the beginning, later rhythms having rests at the end. We then repeat this for $s = 2$ and values of l from 2 to n, and continue in a similar manner until we reach $s = 8$, $l = 8$.

Examples

Here we return to traditional notions in order to exemplify how the sequence begins (ex.18.1), how it processes through the patterns that include Beethoven's 5th (ex.18.2), and how the sequence (five years later) finally ends (ex.18.3).

Table 18.4 Order for performance

Ex.18.1	BWW-[8,0,0](-)--	Two bars rest.
	BWW-[8,1,1](1)-c0 to BWW-[8,1,1](1)-c7	The first rhythm (single quaver at the end) with all 8 possible tunes.
	BWW-[8,1,1](2)-c0 to BWW-[8,1,1](2)-c7	The second rhythm (a single quaver in the penultimate position).
	...	
	BWW-[8,1,1](8)-c0 to BWW-[8,1,1](8)-c7	The last single quaver rhythm (note at the start).
	BWW-[8,1,2](1)-c0 to BWW-[8,1,2](1)-c7 to BWW-[8,1,2](7)-c0 to BWW-[8,1,2](7)-c7	The single crotchet tunes: -note at the end to -note at the beginning.
	...	
	BWW-[8,1,8](1)-c0 to BWW-[8,1,8](1)-c7	End of the first row (single note tunes).
	BWW-[8,2,2](1)-c0 to BWW-[8,2,2](1)-c63 to BWW-[8,2,7](18)-c0 to BWW-[8,2,7](18)-c63	Start two note tunes with two quavers.
	... (Ex.2, B5 in here somewhere)	
Ex.18.3	BWW-[8,8,8](1)-c0 to BWW-[8,8,8](1)-c16777215 (This row alone accounts for one fifth of the tunes!)	8 quavers – The end.

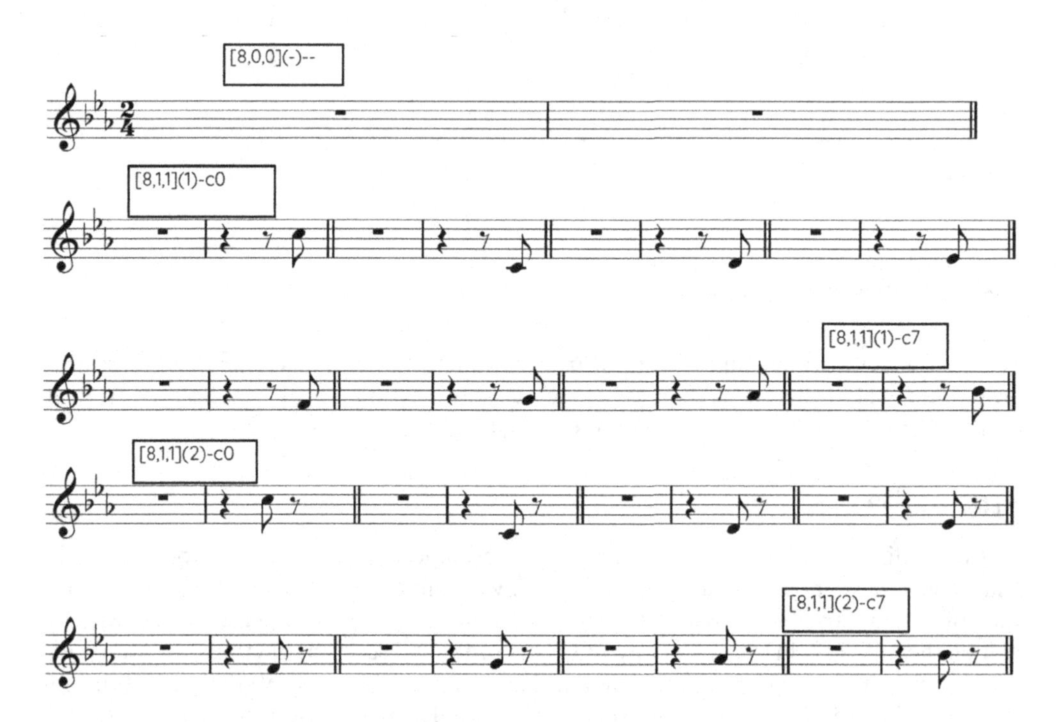

Example 18.1 Sequence start (the first 17 phrases)

Example 18.2 Sequence that includes 'Beethoven's 5th'

Example 18.3 Sequence end (final 10 phrases)

Coda

Our system produces over 83.7 million tunes (83,739,041 to be exact). It would be around 1.3 billion if all 12 semitones were used, 10.2 billion for 2 octaves of 8 notes, and almost 200 billion for 2 octaves with all semitones. Were we to add in greater rhythmic complexity and extend the sequence then it would be in the trillions and beyond. But we have shown that the tunes can be made in a logical system. The BWW system may only account for the first 83.7 million, but the others will follow in time.

19 Deproduction

Robert Wilsmore

The years 1950 to 2100 stand as the great era of popular music. It looked for all the world, at a time in the middle of that period, that the various ages of Western music were over (Baroque, Classical, Romantic, etc.), that the idea that there could be just a few nameable composers as in previous eras had dissipated forever into the vast distributed collaboration of the pop project where the writers in their millions and the songs in their billions, all available to be pulled from the clouds by anyone in the world, and on other worlds at any time (on Mars at least by that point), replaced the individual composer. Genius replaced by Scenius in the acknowledgement of all the contributors. It had seemed that mass communication and distribution had ended the possibility of there ever being observable periods of anything ever again, so disparate and sprawling the world of music had become that finding the lines that would allow the pop object to be identified looked almost impossible. But looking back on it from a distance it had much the same characteristics as other periods of music before it, a creative epoch that exploded, blossomed over a hundred years or so, and then slowed into a long tail whilst creativity found somewhere else to feed and grow. The great production period of popular music was over by the end of its century and a half. Many more decades (centuries even) after that, after the 'long tail', the music was still there, all music was there now waiting to be (re)discovered but not to be created. Not anymore. Following the age of pop there had been a deceleration, a diminuendo, a process of deproduction in which the act and art of production was being erased towards a point where the term itself fell into the archive, used only on occasions in quizzes where contestants had to guess what this archaic term once meant.

The age of pop was aware that in opening up the means of production to just about anyone who could afford the technology, production would increase on an exponential scale leading up to the point where the technology would take over production and produce everything. Back at the start of the 21st century Nicholas Bourriaud wrote that "in our daily lives, the gap that separates production and consumption narrows each day" (2007, p.39). He borrowed the term 'postproduction' from the music and film industries and applied it to art in general where the ready-mades were the building blocks of art for a generation where recycling was not merely a late postmodernist mantra but a worldwide necessity. Bourriaud's postproduction was still production, but production with an awareness of the *always-already* contained within it. But what was eventually to come may have been more disturbing to the pop generations whose identity wrapped itself in the clothing of its music and its artists. Devices, guided by bio and neural feedback, applied the 'right' music to the user that would produce a desired *affect*. Like a drug, it was no longer a choice that one would like the given music but a certainty. The choice not to use these devices was an option, an option some considered as a route to an authentic self (just as the pop age had done in its early stages), whilst others did not feel that need (just as those in the later stages of the pop age had done). The sharing of music was more problematic, finding the optimal

DOI: 10.4324/9781351111959-23

affect for everyone in the group meant that any one individual's affect was not optimal for them. Most devices had an option that allowed individuals in the group to seemingly enjoy a shared experience whilst each was fed something unique to them, allowing them to feel the same level of affect if not the same music. People were then curious as to what others were experiencing and so they could plug in to someone else to get *their* experience as well as their own. That was fun, it was an insight into what moved others and was a great conversation starter. All music was already in existence but these particular devices tended to 're-'create music that already existed, not even needing to access the *already-there* but just creating the already created again (pragmatically this was easier to do than accessing the existing). The 'great copyright fight' that followed the age of pop had long since passed and so was not an issue for anyone anymore, and the absolute deterritorialisation of production was complete by this point. But with the impossibility of *knowing* all that existed, discovering again was as good as creating for those that had the impulse to do so. As is often the case, children's stories carry with them a profound truth disguised in naïve clothing, and a centuries old story by A.A. Milne of a boy and a bear started with a cute and curious creative conundrum of the boy pulling his teddy bear down the stairs:

> Here is Edward Bear, coming downstairs now, bump, bump, bump, on the back of his head, behind Christopher Robin. It is, as far as he knows, the only way of coming downstairs, but sometimes he feels that there really is another way, if only he could stop bumping for a moment and think of it. And then he feels that perhaps there isn't.
>
> (Milne 1926)

If he could stop for a moment then the creative problem-solving bear might come up with the solution that walking down stairs would be a better way of getting from top to bottom. That, of course, is the joke of the situation. Walking down stairs already exists, but he hasn't figured that out yet even though we, the onlooker, know this. And so it became with music once the processes of deproduction were complete. But rediscovering was only one way to access the existing music, just as the idea of walking down stairs was, back then, the common way to descend. Creativity is viewed differently now to then, that's all. Everyone, every musicker at least, could now be seen to be in the position of permanently 'head bumping', searching to find the already-there. There is still talk occasionally of an impending wipeout, an erasure of all music. Perhaps that is what is needed, a clean sheet, a *tabula rasa* but one that has no awareness of a pre-existence, then everything could be new, be created rather than recreated. And it would be no real loss, Beethoven won't know or care that neither he nor his symphonies effectively never existed, and neither would anyone else.

Deproduction was the collective, coproduced, end of the art of production. This was nothing new – art has always ended. Back as far ago as Hegel's lectures on Aesthetics in Berlin in the 1820s, he spoke extensively on the subject of 'the end of art' and this 'end' took several forms. In one form it veered towards purpose, as in *'to what end'* is the activity undertaken when the artist borrows from nature or from the senses, "what the interest or the *End* is which man proposes to himself when he reproduces such a content in the form of works of art" (Hegel 1993, p.46). It could, to Hegel, be as simple as an end in itself for mere entertainment, but that was not a concern of much interest to him, a bit more important was whether it might have a moral impulse, but even more than that was that art was a way that the *Idea*, god-like and self-obsessed, could come to know itself. Did the cave-dwelling bison painter have a different purpose to their art other than to have a bison in the cave in the absence of a real one? That is perhaps different to the start of music recording where it was about having the music in the room in the absence of the musicians (*deus ex machina* through the phonograph in Eisenberg's view). Or so we presumed; who is to say that the cave dweller wanted the same end to their art, that is, to replace the absent bison as best as they could? And besides, bison are big and smelly and take up room, so a painting is better in many ways, hardly a case of "superfluous labour falling short of nature" as Hegel would have it. And who really wants the actual musician in the room? That would just be awkward; one would have to make them a cup of tea and engage in small talk about the weather and ask what they are going to do with the rest of their day and whether they wanted a lift *(and please say 'no')*. The phonograph presented no such demands or dilemmas, and the bison painting required no mucking out. Perhaps this mimeticism was 'more than' rather than 'less than' in relation to the thing itself on a fundamentally pragmatic level. Art for Hegel had the possibility as being an end in itself rather than just being instrumentalised for another purpose but that its part was to move the *Idea* towards self-actualisation, and if art is an instrument with an end outside itself then there may be other ways to get to that end which may dispense with art (which is why it had ended). Given that the absolute *Idea* makes art manifest in sensuous form in order to come to know itself then it is no longer needed once that job is complete or is better fulfilled by another tool such as philosophy or religion. The pantheon of arts of Ancient Greece – architecture, sculpture, painting, music, and poetry – could move towards self-realisation, but after that, where philosophy and religion could progress, "Art, by contrast, cannot reach such a climax, but peters out in reflection and irony" (Woods in Hegel 1993, p.xxxi). Art then, in Hegel's time (and according to Hegel) was already in decline, its deproduction in progress and its long tail on an inevitable journey forever thinning as it goes. At least, as far he could see at the time. It might be possible, in an unknown yet-to-come, that it may have an end 'again' placed in its future, an opportunity to be *becoming* once more, and, as we had grown used to, endings can be annulled. And it was 150 years or so after Hegel's lectures before art ended again having contemplated 'art after the end of art'. But art after the end of art is different to art after the end of production. Music, in Hegel's time of realising the Romantic idea, "liberates the ideal content from its immersion in matter ... [and] finds utterance for the heart with its whole gamut of feelings and passion" (1993, p.95). To *know* how we feel, we need to have feelings. And Hegel leaves no doubt as to the instrumentalisation of art at the end of his lectures:

> And, therefore, what the particular arts realise in individual works of art are according to their abstract conception simply the universal types which constitute the self-unfolding Idea of beauty. It is the external realisation of this Idea that the wide Pantheon of art is being erected, whose architect and builder is the spirit of beauty as it awakens to self-knowledge, and to complete which the history of the world will need its evolution of ages.
>
> (1993, p.97)

Art, already past its ancient glory, still processes through ages. Eva Guelen in *Hegel Without End,* wrote that "Whatever beautiful art is, it perishes and dissolves between birth and rebirth. The end of art is the process of art, and it passes not from art to philosophy (or vice versa) but passes away between art and philosophy" (2006, p.22). After the classical art of ancient Greece where "the possibilities of art are completed, and by the same token, exhausted. Almost everything in this relation depends upon the concept of 'recollection'" (2006, p.23). Art after the end of art is art before its end *again*, it is *'re'*-art, "rediscovery, recognition, and rebirth", the past contained, sublated, in the present.

The pop age, driven by its explosion of technological production and dissemination, allowed the *Idea* of beauty to come to know itself in the vast oneness of its seemingly billions of songs, rather than its evolution of singular symphonies. But the *Idea* was at odds with the concept of the singularity of 'The Song of a Thousand Songs'. Hegel had an Idea, Toast on the other hand, had no Idea. If reduced to the 'simple' then Toast simply *was* without need of an absolute, without an inaccessible being *needing* self-knowledge. Hegel's beauty is so 'needy', does it really need so many songs, so many art works, about itself? Perhaps that's where the classical tale and images of Narcissus denote that period as being the end of art, beauty finally looking at itself in a moment of self-satisfied contemplation of its own perfection and completion. A moment of self-consumption – the very thing pop's future was supposed to have held (it had been prophesied). As much as the Deleuzoguattarian exposing of continuity over separation of the songs held for both with regard to the repositioning (or denial) of the individual underneath, the latter was not a simple revealing of universal types but rather the revealing of the plane of consistency by tearing off the markers of segmentation. Sally Macarthur wrote of the Deleuzian position "When the sonic object opens to the plane of immanence the sonic object fades. The distinction between the listener and the sound object, the listener and the hearer, and listening and vision vanish" (Macarthur, Lockhead, and Shaw 2019, p.176).

There was concern, in the age of pop, that the accessible technology for production and dissemination homogenised users into pseudo-individualisation through micro-differentiations (Adorno had thought so), which was nothing beyond worthless reflections of a universal form, and it was true that monetisation in the *milieu* was hard (the streams pouring from the clouds brought little money with them). The *Idea* could take nothing from it, and whereas it may have gained something of use from the art of recording that had come long after the art of ancient Greece (recording being a nugget of knowledge found in the grits of the long tail), there was nothing of value left once deproduction had set in. And the devices that came later (the ones with the guaranteed *affect*) were fought against by those threatened by the erosion of self, but their defence was relatively short lived. Again, the *Idea* was not impressed, through the law of diminishing returns it had decided it was no longer worth the effort of sorting through the grits. Nihilistic and prophetic, darkened and cancelled in acceleration, Nick Land foreshadowed that "Mass computer commoditization de-differentiates consumption and investment, triggering cultural micro-engineering waves that dissociate theopolitical action into machinic hybridites, amongst increasingly dysfunctional defensive convulsions" (Land 2011, p.397). Pockets of resistance, brief spasms of a light blinking until dark forever, weakened and relaxed into the plane, becoming contingent once more, manifestations in devices arising only by *taking* from ($n-1$), not *growing* from ($n+1$), the planes, and even then barely managing to emerge, they eventually rippled into stillness.

As with art in general, the end of music production was not the *death* of music. The activity of production, the art of production itself may have ended but not music or the recording. Over a century and a half after Hegel's lectures, Arthur Canto wrote in *After the End of Art* (1997) that the end of art was not the same as the death of art:

> It was not my view that there would be no more art, which '*death*' certainly implies, but that whatever art there was to be would be made without benefit of a reassuring sort of narrative in which it was seen as the appropriate next stage in history. What had come to an end was that narrative but not the subject of that narrative.
>
> (Canto 1997, p.4)

The era of the age of pop was, in the 1980s in which 'art after the end' was becoming a familiar discourse, coming into its own in the destruction of the single genius and the rise of the many and dismantling the grand narrative, only in the end (by which we mean as a term of hindsight, concretising the apparent contemporary fluidity through reflection that retroactively solidifies the past) to be seen as another epoch, similar but different to those that came before it. Canto wrote that the end of one narrative was not 'the' end, as "life really begins when the story comes to an end" (1997, p.4), and that "The claim that art is ended is really a claim about the future – not that there will be no more art, but that such art as there will be is art after the end of art, or, as I have already termed it, *post-historical art*" (1997, p.43). Such writings about the future are not "predicting but prophesying". In truth, we compare the now to history, but we also compare the now to the future, even if we are speculating on the turn of events in order to provide the model by which to make the comparison. The *Idea* is, for us, one existing model by which to view the now. If we decide that it is not the *Idea* and its self-knowledge that is the reality, but instead that *Creativity* is the real god, then Creativity in its self-realisation finds other fertile ground when it is finished with one field, tossing aside the eras, artists, genres and styles as it goes (it may even toss aside *Idea* once it has been used for its own purposes). And if production of music was one of those then its tossing aside, its deproduction, was another bloom and decay, the irreversible natural process of entropy where dissipation becomes so wide into the plane it is rendered beyond invisible and into non-existence. And yet production (in the form of record production) had such humble beginnings, wax disks and phonographs, then later the lone figure sat at a large mixing desk strewn with knobs and faders, with all that has been captured from the virtuosic performers, creative artists, and genius songwriters at their fingertips. After that it was no longer a solitary act but an act of coproduction, in twos, in threes, in tens, in millions. After that the technology took over and finished the job of producing everything. After that, here and now, at what looks like another end, we await with excitement as to how this end will be annulled and, in some future reflection, possibly be revealed as the continuous collaboration of coproducers.

Glossary of Terms

Complementary Coproduction Productions characterised by division of labour according to the different and 'complementary' skills of the contributors (after Vera John-Steiner).

Consolidation Points Framework (CPF) Points on a linear perspective of the *Production Habitus* where focus on particular 'non-sonic' elements is given attention.

Copyleft In binary opposition to copyright, copyleft is the practice that ideas, intellectual property, etc. can be used and distributed freely, and that derivations from this will similarly be allowed to be used and distributed freely.

Deproduction The process by which society moves towards the position where there is nothing left to produce and hence the act of and the term 'production' become obsolete.

Distributed Coproduction Productions characterised by widespread and often voluntary contributions to a project where parameters and hierarchies are not well defined or are absent (after Vera John-Steiner). 'The Song of a Thousand Songs' (see Toast theory) is an extreme example of this mode of coproduction.

Familial Coproduction Productions characterised by established teams where roles and skills easily interchange and cross over in a fluid and implicitly accepted manner (after Vera John-Steiner).

Group Think A term coined by Irving Janis that describes the ability for a group to arrive at a conclusion or decision that is irrational, often occurring where the overriding desire of the team to maintain harmony takes priority over 'good' decision making.

Head Bumping The act of trying to invent or discover for oneself something that others already know but you don't (after Winnie-the-Pooh).

Holon A term coined by Arthur Koestler to describe something that is a whole in itself as well as being a part of another whole.

Integrative Coproduction Productions characterised by transformative collaborations and where individual contribution is no longer identifiable (after Vera John-Steiner).

Internal Coproduction The recognition of the different influences that are at play in the production process of the individual. It recognises the role of the 'other' within the self in the production process.

Organising Genius Bennis and Biederman's study that challenges the 'myth of the great man' and reveals the success of the creative team rather than that of the individual.

Production Habitus Derived from Bourdieu's concept of habitus, the *Production Habitus* delineates the socially ingrained behaviours, environment, accepted norms, capital (financial, social, and cultural) and assumed expectations that act upon music production.

Rhizomatics The concept based on the rhizome, developed by Deleuze and Guattari in their book *A Thousand Plateaus*, rethinks a 'tree-like' conception of the world into a non-linear, non-hierarchical and everchanging map that one may enter and exit at any point, taking 'lines of flight' along the way. It establishes connection over disconnection, refuting ends through continual deferral to the conjunctive 'and'.

Scenius A term coined by Brian Eno to express how genius is not isolated to great individuals but necessarily includes the wider community and environment, presumably a portmanteau of the words 'scene' and 'genius' to connect the two together.

Toast Theory The idea that there is only *one song* in the world that everyone contributes to, which is called 'The Song of a Thousand Songs' (or 'Toast' for short). Following concepts of Deleuze and Guattari, it questions the validity of endings of 'individual' songs and in doing so emphasises the positivity of connection and collaboration over the negativity of separation that prioritises ownership leading to accusations of theft and plagiarism.

References

Abbey Road (2020) 'Studio Two' [Online]. Available from: www.abbeyroad.com/studio-two [Accessed 20 November 2020].

Adorno, T. and Simpson, G. (1941) 'On Popular Music'. Originally published in *Studies in Philosophy and Social Science*, New York, Institute of Social Research, 1941, IX, 17–48.

Albiez, S. and Pattie, D. (eds) (2016) *Brian Eno. Oblique Music*. London and New York, Bloomsbury.

Alexander, J.S. (2007) 'Production Construction', *Mix,* 31 (8), pp.32.

Alix, C., Dobson, E., and Wilsmore, R. (2010) *Collaborative Arts Practices in HE: Mapping and Developing Pedagogical Models.* Palatine, The Higher Education Academy.

Allain, R. (2015) 'How Many Different Songs Can There Be?', *Wired* [Online]. Available from www.wired.com/2015/03/many-different-songs-can/ [Accessed 3 November 2018].

Amabile, T.M. (1983) 'The Social Psychology of Creativity: A Componential Conceptualization', *Journal of Personality and Social Psychology,* 45 (2), pp.357–376.

Amabile, T.M. (1996) *Creativity in Context: Update to "The Social Psychology of Creativity"*. Boulder, CO, Westview Press.

Amabile, T.M. (2013) 'Componential Theory of Creativity' in Kessler, E. (ed.) *Encyclopedia of Management Theory*. Thousand Oaks, SAGE Publications Ltd, pp.135–139.

Amabile, T.M. and Pillemer, J. (2012) 'Perspectives on the Social Psychology of Creativity', *The Journal of Creative Behavior,* 46 (1), pp.3–15.

Annand, S. (2010) *The Half* [Online]. Available from www.simonannand.com/the-half [Accessed 8 September 2019].

Arnold, R. (2003) *Empathic Intelligence: The Phenomenon of Intersubjective Engagement*. Conference paper First International Conference on Pedagogies and Learning. University of Southern Queensland.

Arnstein, S. (1969) 'A Ladder of Citizen Participation', *Journal of the American Planning Association*, 35 (4), pp.216–224.

Auslander, P. (1999) *Liveness: Performance in a Mediatised Culture*. London and New York, Routledge.

Backer, S. (2018) SPOT [Online]. Available from https://daily.redbullmusicacademy.com/2018/11/behind-the-sound-of-american-punk [Accessed 30 July 2019].

Badiou, A. (2007) *Being and Event*. Trans. Oliver Feltham. London and New York, Continuum.

Bargfrede, A. (2017) *Music Law in the Digital Age: Copyright Essentials for Today's Music Business*. 2nd ed. Boston, MA, Berklee Press.

Barker, H, and Taylor, Y. (2007) *Faking It: The Quest for Authenticity in Popular Music.* London, Faber and Faber Limited.

Barsanti, S. (2018) 'Ed Sheeran Is Getting Sued Again' [Online]. Available from www.avclub.com/ed-sheeran-is-getting-sued-again-1821997610 [Accessed 18 August 2018].

Barthes, R. (1977) *Image – Music – Text*. Trans. Stephen Heath. New York, Hill and Wang.

Bartleet, L. (2018) 'Radiohead v Lana Del Rey: A Comprehensive Timeline of the Most Confusing Music "Lawsuit" in Recent Music History', *NME* [Online]. Available from www.nme.com/blogs/nme-blogs/radiohead-v-lana-del-rey-creep-get-free-lawsuit-timeline-2216897 [Accessed 18 January 2018].

Baudrillard, J. (1994 [1981]) *Simulacra and Simulation*. Trans. Sheila Glaser. Ann Arbor, MI, University of Michigan Press.

BBC (2018) 'Lana Del Rey Says Radiohead Are Suing Her', BBC Music reporter Mark Savage [Online]. Available from www.bbc.co.uk/news/entertainment-arts-42602900 [Accessed 18 January 2018].

Belbin, M. (N.D) *Team Roles* [Online]. Available from www.belbin.com/about/belbin-team-roles/ [Accessed 18 January 2018].

Bennett, J. (2011) 'Collaborative Songwriting: The Ontology of Negotiated Creativity in Popular Music Studio Practice', *Journal on the Art of Record Production*, 5.

Bennett, J. (2018) 'Songwriting, Digital Audio Workstations, and the Internet', *The Oxford Handbook of the Creative Process in Music*, 1st ed. Oxford University Press.

Bennis, W. and Biederman, P. (1997) *Organizing Genius: The Secrets of Creative Collaboration*. Cambridge, MA, Perseus Books.

Beresford, P. (2013) *Beyond the Usual Suspects*. London, ShapingOurLives.org.

Bevan, D. (2012) 'Bon Iver Announces Remix Contest: The Stems Project' [Online]. Available from www.spin.com/2012/08/bon-iver-announces-remix-contest-stems-project/ [Accessed 4 November 2020].

Billboard (2017) 'Foo Fighters Rickroll, Calvin Harris Blasts the Hits at Japan's Summer Sonic' [Online]. Available from www.billboard.com/articles/news/festivals/7940970/foo-fighters-rickroll-calvin-harris-japan-summer-sonic. [Accessed 28 September 2018].

BMI (2017) *How I Wrote That Song 2017: Lauren Christy on Staying Creative & Spotting Genius*. BMI published YouTube 17 February 2017.

Bourdieu, P. (1977) *Outline of a Theory of Practice*. Cambridge, Cambridge University Press.

Bourdieu, P. (1991) *Language & Symbolic Power*. Cambridge, Polity Press.

Bourdieu, P. (2010 [1984]) *Distinction: A Social Critique of the Judgment of Taste*. London and New York, Routledge.

Bourriaud, N. (2007) *Postproduction*. New York, Lukas and Sternberg.

Bovaird, T. (2007) 'Beyond Engagement and Participation: User and Community Co-production of Public Services', *Public Administration Review*, 67 (5), pp.846–860.

Braude, S.E. (1995) *First Person Plural: Multiple Personality and the Philosophy of Mind*. Revised edition. Maryland and London, Rowman & Littlefield.

Buchanon, I. and Swiboda, M. (eds) (2004) *Deleuze and Music*. Edinburgh, Edinburgh University Press.

Bullingham, L. and Vasconcelos, A.C. (2013) 'The Presentation of Self in the Online World: Goffman and the Study of Online Identities', *Journal of Information Science*, 39 (1), pp.101–112.

Burgess, R.J. (2002) *The Art of Music Production*. New York and London, Omnibus Press.

Burgess, R.J. (2004) 'Interview with Lauren Christy of The Matrix', *The Journal on the Art of Record Production*, issue 5, July 2011.

Burgess, R.J. (2013) *The Art of Music Production*, 4th edition. Oxford, OUP.

Burkholder, J. P. (1985) 'Quotation and Emulation: Charles Ives's Uses of His Models', *The Musical Quarterly* 71 (1), pp.1–26.

Burkholder, J. P. (1994) 'The Uses of Existing Music: Musical Borrowing as a Field', *Notes* 50 (3), 851–870.

Burrows, J. (2016) *Rhizo-Memetic Art: The Production and Curation of Transdisciplinary Performance*. PhD Thesis, Edge Hill University.

Buskin, R. (2006) 'The Matrix', *Sound on Sound*, April 2006.

Butler, M. (2003) 'Taking It Seriously: Intertextuality and Authenticity, Two Covers by the Pet Shop Boys', *Popular Music*, 22 (1), pp.1–19.

Campbell, J. (2008) *The Hero with A Thousand Faces*. California, New World Library.

Canto, A. C. (1997) *After the End of Art: Contemporary Art and the Pale of History*. New Jersey and Oxford, Princeton University Press.

Cauty, J. and Drummond, B. (2001) *The Manual (How to Have a Number One the Easy Way)*. London, Ellipsis.

Chazelle, D. (2016) *La La Land*. Film script.

Clark, N. (2020) 'Hours About That – Daily walk or run should be for a maximum of one hour and near home during coronavirus lockdown, Michael Gove says' [Online]. Available from www.thesun.co.uk/news/11284289/daily-walk-or-run-maximum-one-hour-coronavirus/ [Accessed 25 Aug 2020].

Cosgrove, S. (2016) *Detroit 67: The Year that Changed Soul*. Edinburgh, Polygon.

Costa, C. and Murphy, M. (2015) *Bourdieu and the Application of Habitus Across the Social Sciences*. London, Palgrave Macmillan.

Cox, B., Ince, R., and Feacham, A. (2017) *How to Build a Universe*. London, Williams Collins Books.

Commuter Films (2021) 'Full Service Video Production Company' [Online]. Available from www.commuterfilms.co.uk/ [Accessed 16 August 2021].

Crossman, A. (2019) 'The Presentation of Self in Everyday Life' [Online]. Available from www.thoughtco.com/the-presentation-of-self-in-everyday-life-3026754 [Accessed 6 September 2019].

Csikszentmihalyi, M. (1996) Creativity! Flow and the Psychology of Discovery and Invention. New York, Harper Collins.

Csikszentmihalyi, M. (2013) *Creativity: The Psychology of Discovery and Invention*. New York, Harper Perennial Modern Classics.

Davies, C. (2017) 'X Factor Sees Lowest Viewing Figures' [Online]. Available from www.theguardian.com/tv-and-radio/2017/sep/03/x-factor-sees-lowest-viewing-figures-since-show-began. Guardian Newspaper [Accessed 16 August 2018].

Dawkins, R. (2016) *The Selfish Gene*. New York, Oxford University Press.

Deleuze, G. (1988) *Spinoza: Practical Philosophy*. Trans. Robert Hurley. San Francisco, City Light Books.

Deleuze, G. (2010) *Difference and Repetition*. Trans. P. Patton. London and New York, Continuum.

Deleuze, G. and Guattari, F. (1987) *A Thousand Plateaus: Capitalism and Schizophrenia*. Trans. Brian Massumi. Minneapolis, MN, University of Minnesota Press.

Deleuze, G. and Guattari, F. (2011) *What Is Philosophy?* Trans. H. Tomlinson and G. Burchill. London and New York, Verso.

Diehl, M. (2010) 'It's a Joni Mitchell concert, sans Joni' [Online]. *Los Angeles Times*. Available from http://articles.latimes.com/2010/apr/22/entertainment/la-et-jonimitchell-20100422/3 [Accessed 22 February 2013].

Dimaggio, G. and Stiles, W.B. (2007) 'Psychotherapy in Light of Internal Multiplicity', *Journal of Clinical Psychology,* 63 (2), pp.119–127.

Edwards, M. (N.D.) *A Brief History of Holons* [Online]. Available from www.integralworld.net/edwards13.html [Accessed 20 August 2019].

Eisenberg, E. (2005 [1987]) *The Recording Angel: Music, Records and Culture from Aristotle to Zappa*. New Haven, CT, Yale University Press.

Eisentein, S. (2010) *Writings 1934–1947: Sergei Eisenstein Selected Works, volume 3*. 1st ed. London, New York, I.B. Taurus.

Eno, B. (1991) 'Bringing Up Baby', *Rolling Stone,* No. 618.

Eno, B. (2009) 'Brian Eno Speaking at the Sydney Luminous Festival' [Online]. Available from www.synthtopia.com/content/2009/07/09/brian-eno-on-genius-and-scenius/ [Accessed 19 February 2021].

Eno, B. (2015) *Brian Eno Speaking at 'Basic Income: How Do We Get There?'* Basic Income UK meet-up at St Clements Church Kings Square, London, 3 December 2015.

Escudero, J.A. (2014) 'Heidegger, on Selfhood', *American International Journal of Contemporary Research,* 4 (2), pp.6–17.

Farr M., Davies R., Davies P., Bagnall D., Brangan E., and Andrews H. (2020) *A Map of Resources for Co-producing Research in Health and Social Care.* National Institute for Health Research (NIHR) ARC West and People in Health West of England; University of Bristol and University of West of England. Version 1.2, May 2020.

Foucault, M. (2011) 'What Is an Author?', *The Continental Aesthetics Reader*, 2nd ed. Abingdon, Routledge, pp. 525–539.

Frith, S. (1996) *Performing Rites: On the Value of Popular Music*. Oxford, OUP.

Frith, S. and Goodwin, A. (eds) (1990) *On Record: Rock, Pop and the Written Word.* London, Routledge.

Frith, S. and Zagorski-Thomas, S (eds) (2012) The Art of Record Production. Surrey and Burlington, VT, Ashgate.

Gallagher, S. and Radden, J. (2011) *Multiple Selves.* Oxford, OUP.

Garcês, S., Pocinho, M., Jesus, S.N. and Viseu, J. (2016) 'The Impact of the Creative Environment on the Creative Person, Process, and Product', *Revista Avaliação Psicológica,* 15 (2), pp.169–176.

Gaventa, J. (2003) *Power after Lukes: An Overview of Theories of Power Since Lukes and their Application to Development.* Brighton, Participation Group, Institute of Development Studies.

Giddens, A. (1991) *The Consequences of Modernity*. Oxford, Blackwell Publishing.

Glynos, J. and Speed, E. (2012) 'Varieties of Co-Production in Public Services: Time Banks in a UK Health Policy Context', *Critical Policy Studies*, 6, pp.402–433.

Goffman, E. (1990) *The Presentation of Self in Everyday Life.* London, Penguin.

Gormely, I. (2012) 'Glen Hansard Goes Solo But Says Swell Season Could Happen Again in the Future' [Online] Spinner. Available from www.spinner.com/2012/05/30/glen-hansard-solo-swell-season/ [Accessed 6 March 2013].

Gracyk, T. (1996) *Rhythm and Noise: An Aesthetics of Rock.* Durham, NC, Duke University Press.

Green, C. (2001) *The Third Hand: Collaboration in Art from Conceptualism to Postmodernism.* Minneapolis, MN, University of Minnesota Press.

Grohl, D. (2013) *On American Idol.* Facebook comment, 3 May 2013. Available from www.facebook.com/permalink.php?story_fbid=606972795981986&id=239510859394850 [Accessed 12 October 2018].

Grout, D.J. (1988) *A History of Western Music*, 4th ed. London, J.M. Dent & Sons.

Gruber, H.E. and Davis, S.N. (1988). *Inching Our Way Up Mount Olympus: The Evolving-Systems Approach to Creative Thinking.* New York, NY, Cambridge University Press.

Guelen, E. (2006) *The End of Art: Readings in a Rumour after Hegel.* Trans. James McFarland. California, Stanford University Press.

Harding. P. (2010) *PWL From the Factory Floor.* London, Cherry Red Books.

Harding, P. (2020) *Pop Music Production: Manufactured Pop and Boy Bands of the 1990s.* Oxon and New York, Routledge.

Harman, G. (2019) *Object Oriented Ontology.* London, Pelican Books.

Hegel, G, W, F. (1993) *Introductory Lectures on Aesthetics.* Trans. Bernard Bosanquet. London and New York, Penguin Books.

Heidegger, M. (1967) *Being and Time.* Trans. Macquarrie, J. and Robinson, E. Cambridge, MA, and Oxford, Blackwell.

Helpmusicians.org (2019) 'Smoke Rainbows Album in Aid of Music Minds Matter Announced' [Online]. Available from www.helpmusicians.org.uk [Accessed 16 December 2020].

Hepworth-Sawyer, R. and Golding, C. (2011) *What Is Music Production? A Producer's Guide: The Role, the People, the Process.* London and New York, Taylor & Francis.

Hepworth-Sawyer, R. and Hodgson, J. (2017) *Perspectives on Music Production: Mixing Music.* New York, Routledge.

Heylighen, F. (2009) *What Makes a Meme Successful? Selection Criteria for Cultural Evolution.* Conference proceedings from the 15th Int. Congress on Cybernetics. [Online]. Available from http://cogprints.org/1132/1/MemeticsNamur.html. [Accessed 16 March 2017].

Hickey, G., Brearley, S., Coldham, T., Denegri, S., Green, G., Staniszewska, S., Tembo, D., Torok, K., and Turner, K. (2018) *Guidance on Co-producing a Research Project.* Southampton: INVOLVE.

Hillborg, A. (N.D.) BBC Proms [Online]. Available from www.hillborg.musikelit.nu [Accessed 12 February 2004].

Hogan, B. (2010) 'The Presentation of Self in the Age of Social Media: Distinguishing Performances and Exhibitions Online', *Bulletin of Science, Technology & Society,* 30 (6), pp.377–386.

Holland, J. (2015) 'WholeWorldBand' [Online]. Available from www.idaireland.com/newsroom/whole-world-band-takes-world. [Accessed 24 August 2020].

Horne, M., Khan, H., and Corrigan P. (2013) *People Powered Health: Health for People, By People and With People.* London, Nesta.

Howard, D. (2004) *Sonic Alchemy: Visionary Music Producers and Their Maverick Recordings.* New York, Hal Leonard.

Hughes, A. (1979) 'Dimitri Tiomkin Dies', *The New York Times*, 14 November 1979.

IFAI (2014) Conference paper at *The International Festival for Artistic Innovation.* Leeds, UK, Leeds College of Music.

Ignatow, G. (2009) 'Why the Sociology of Morality Needs Bourdieu's Habitus', *Sociological Inquiry,* 79 (1), pp.98–114.

Jenkins, R. (2010) 'The 21st-century Interaction Order', in *The Contemporary Goffman*. New London, Routledge, pp.271–288.

Johnson, C. (2020) 'Silver Glass: A Live Solo Version with Chris Johnson from Mostly Autumn/ Halo Blind' [Online]. Available from www.youtube.com/watch?v=I5ZmK7qwRyU [Accessed 18 August 2020].

John-Steiner, V. (2000) Creative Collaboration. New York, Oxford University Press.

Jordanous, A. (2016) 'Four Perspectives on Computational Creativity in Theory and in Practice', *Connection Science,* 28 (2), pp.194–216.

Kawashima, D. (2006) 'Writing/Producing Trio The Matrix Break Through With Hits For Avril Lavigne and Hilary Duff' [Online]. Available from www.songwriteruniverse.com/matrix.htm [Accessed 17 September 2021].

Kerrigan, S. (2013) 'Accommodating Creative Documentary Practice within a Revised Systems Model of Creativity', *Journal of Media Practice,* 14 (2), pp.111–127.

Kelly, A. F. (2012) 'Cover Songs: Does Validity Equal Authenticity?' [Online]. Mungbeing, Issue 15. Available from www.mungbeing.com/issue_15.html?page=35#1221 [Accessed 9 November 2012].

Koestler, A. (1975) *The Ghost in the Machine*. London, Pan Books.

Ladbrokes (2017) 'Wham! Favourites to Land This Year's Christmas Number One' [Online]. Available from http://news.ladbrokes.com/novelty/wham-favourites-to-land-this-years-christmas-number-one.html [Accessed 16 September 2018].

Lambert, N. and Carr, S. (2018) 'Outside the Original Remit': Co-Production in UK Mental Health Research, Lessons from the Field', *Int J Mental Health Nurse,* 27, pp.1273–1281. doi:10.1111/ inm.12499.

Lamont, M. (1992) *Money, Morals, and Manners: The Culture of the French and the American Upper-Middle Class*. Chicago, University of Chicago Press.

Lana Del Rey (2018) *It's True About the Lawsuit*. [Twitter post] posted 7 January 2018 [Accessed 8 January 2018].

Land, N. (2011) *Fanged Noumena: Collected Writings 1987–2007*. Falmouth and New York, Sequence Press.

Lashua, B. and Thompson, P. (2016) 'Producing Music, Producing Myth? Creativity in Recording Studios', *Journal of the International Association for the Study of Popular Music,* 6, pp.70–90.

Latour, B. (2007) *Reassembling the Social*. New York, Oxford University Press.

Latour, B. (2017) 'On Actor-Network Theory. A Few Clarifications, Plus More Than a Few Complications', *Philosophical Literary Journal Logos,* 27 (1), pp.173–197.

Leslie, K. (2012) 'Glen Hansard: The Tortured Romantic Returns to Atlanta' [Online]. The Atlanta Journal-Constitution. Available from www.ajc.com/news/entertainment/music/glen-hansard-the-tortured-romantic-returns-to-atla/nSBnd/ [Accessed 6 March 2013].

Lewis, A., King, T., Herbert, L., and Repper, J. (2017) *Co-production – Sharing Our Experiences, Reflecting on Our Learning*. Nottingham, IMROC.

LIMTEC (2003) *Leeds International Music Technology Education Conference*. Leeds College of Music.

Liu, D., Chen, X., and Yao, X. (2011) 'From Autonomy to Creativity: A Multilevel Investigation of the Mediating Role of Harmonious Passion', *Journal of Applied Psychology,* 96 (2), pp.294–309.

Llewellyn, M. (2012) 'Theordor Adorno' [Online] Available from www.aber.ac.uk/media/Students/ mml9701.html [Accessed 13 November 2012].

Lovelock, J. (2016) *Gaia*. Oxford and New York, Oxford University Press.

Locock, L., Boylan, A.M., Snow, R., and Staniszewska, S. (2017) 'The Power of Symbolic Capital in Patient and Public Involvement in Health Research', *Health Expert,* 20 (5), pp.836–844. doi: 10.1111/hex.12519. Epub 2016 Nov 24. PMID: 27885770; PMCID: PMC5600212.

Macarthur, S., Lockhead, J., and Shaw, J. (eds) (2019) *Music's Immanent Future: the Deleuzian Turn in Music Studies*. Oxford, NY, Routledge.

Marketwired.com (2009) Uplaya [Online] Available from www.marketwired.com/press-release/music-intelligence-solutions-inc-announces-launch-uplayacom-analyze-hit-potential-promote-1222095. htm [Accessed 20 September 2018].

Marrington, M. (2011) 'Experiencing Musical Composition in the DAW: The Software Interface as Mediator of the Musical Idea', *Journal on the Art of Record Production,* (5).

Marrington, M. (2016) 'Paradigms of Music Software Interface Design and Musical Creativity' in Hepworth-Sawyer, R. Hodgson, J. Paterson, J.L. and Toulson, R. (eds) *Innovation in Music II.* Shoreham-by-sea, Future Technology Press, pp.52–63.

Marshall, A. (2018) 'Music Minds Matter: New Helpline Aims to Lend an Ear to Musicians in Need' Independent [Online], Monday 5 February. Available from www.independent.co.uk/arts-entertainment/music/music-mind-matters-helpline-musicians-mental-health-depression-debt-talk-charity-a8184931.html [Accessed 16 December 2020].

Martin, G. and Hornsby, J. (1979) *All You Need is Ears.* New York, St Martin's Press.

Massey, H. (2000) *Behind the Glass Vol.1: Top Record Producers Tell How They Crafted the Hits.* San Francisco, CA, Backbeat Books.

Massey, H. (2009) *Behind the Glass Vol.2: Daniel Lanois, T-Bone Burnett, Mark Ronson, Hugh Padgham and Many More.* San Francisco, CA, Backbeat Books.

McCabe, C. (1984) *Artistic Collaboration in the Twentieth Century.* Washington, DC, Smithsonian Institution Press.

McCoy, J.M. and Evans, G.W. (2002) 'The Potential Role of the Physical Environment in Fostering Creativity', *Creativity Research Journal,* 14 (3–4), pp.409–426.

McGee, A. (2007) *Must Rock be 4 Real? The Guardian.* [Online], 2 October 2007. Available from www.theguardian.com/music/musicblog/2007/oct/02/mustrockbe4real [Accessed 5 September 2019].

McIntyre, P. (2008a) 'Creativity and Cultural Production: An Interdisciplinary Approach to Understanding Creativity Through an Ethnographic Study of Songwriting', *Cultural Science Journal,* 1 (2).

McIntyre, P. (2008b) 'The Systems Model of Creativity: Analyzing the Distribution of Power in the Studio', *Journal on the Art of Record Production,* (3).

McIntyre, P. (2008c) 'Creativity and Cultural Production: A Study of Contemporary Western Popular Music Songwriting', *Creativity Research Journal,* 20 (1), p.40.

McIntyre, P. (2009). 'Rethinking Communication, Creativity and Cultural Production: Outlining Issues for Media Practice', *Australian and New Zealand Communications Association Annual Conference, & NBSP;* 8–10 July 2009.

McIntyre, P. (2011) 'Systemic Creativity: The Partnership of John Lennon and Paul McCartney', *Musicology Australia,* 33 (2), p.241.

McIntyre, P. (2012) 'Rethinking Creativity: Record Production and the Systems Model', in Frith, S. and Zagorski-Thomas, S. (eds) (2012) *The Art of Record Production: An Introductory Reader for a New Academic Field.* Surrey and Burlington, VT, Ashgate, pp.149–161.

McIntyre, P. (2019) 'Taking Creativity Seriously: Developing as a Researcher and Teacher of Songwriting', *Journal of Popular Music Education,* 3 (1), pp.67–85.

McLamore, A. (2012) 'Musical Performance and Audiences – Performance Considerations' [Online]. Available from http://science.jrank.org/pages/10335/Musical-Performance-Audiences-Performance-Considerations.html [Accessed 9 November 2012].

Metzger, P. (2016) 'The Millennial Whoop: A Glorious Obsession with the Melodic Alternation between the Fifth and the Third' [Online]. Available from: https://thepatterning.com/2016/08/20/the-millennial-whoop-a-glorious-obsession-with-the-melodic-alternation-between-the-fifth-and-the-third/ [Accessed 5 May 2017].

Michaels, M. (2017) *Actor-Network Theory. Trials, Trails and Translations.* LA and London, Sage.

Miranda, E. (2012) 'On Computer-aided Composition, Musical Creativity and Brain Asymmetry' in Collins, D. (ed.) *The Act of Musical Composition: Studies in the Creative Process.* SEMPRE Studies in the Psychology of Music. Farnham, Ashgate Publishing, Ltd, pp. 215–231.

Moorefield, V. (2005) *The Producer as Composer: Shaping the Sounds of Popular Music.* Cambridge, MA, MIT Press.

Middleton, R. (1990) *Studying Popular Music.* Buckingham and Philadelphia, Oxford University Press.

Milne, A.A. (1926) *Winnie-the-Pooh.* London, Methuen.

Moore, A. (2002) 'Authenticity as Authentication', *Popular Music,* 21 (2), pp.209–233.

Moore, A. (2012) *Song Means: Analysing and Interpreting Recorded Popular Song.* Farnham, Ashgate.
MrSuicideSheep (2013) 'Zeds Dead – By Your Side' [Online]. Available from https://www.youtube.com/watch?v=buY2W8rQwoQ [Accessed 25 August 2020].
Mulch, S.M. (2004) 'Interview' [Online]. Available from https://ink19.com/2004/12/magazine/interviews/the-matrix [Accessed 5 October 2018].
Mullin, K. (2012) 'Glen Hansard' [Online]. *Time Out: Sydney*. Available from www.au.timeout.com/sydney/music/features/11754/glen-hansard [Accessed 6 March 2013].
Music Minds Matter (2020) Available from www.musicmindsmatter.org.uk [Accessed 16 December 2020].
Navarro, Z. (2006) 'In Search of a Cultural Interpretation of Power: The Contribution of Pierre Bourdieu', *IDS Bulletin,* 37 (6), pp.11–22.
Needham, C. and Carr, S. (2009) *Co-production: An Emerging Evidence Base for Adult Social Care Transformation.* London, Social Care Institute for Excellence.
Nesta (2013) *People Powered Health Co-production Catalogue.* www.nesta.org.uk/report/co-production-catalogue/ [Accessed 27 November 2020].
O'Hare, C. (2007) *The Secrets of the Joshua Tree*. Hot Press. Available from www.u2songs.com. [Accessed 25 October 2019].
Oliver, K., Kothari, A., and Mays, N. (2019) 'The Dark Side of Coproduction: Do the Costs Outweigh the Benefits for Health Research?' *Health Res Policy Syst.,* 17(1), p.33.
Oswald, V. (2019) *Katy Perry: Purposeful Pop Icon*. New York, Lucent Press.
Pandora (N.D.) *Music Genome Project* [Online]. Available from www.marketwired.com/press-release/music-intelligence-solutions-inc-announces-launch-uplayacom-analyze-hit-potential-promote-1222095.htm [Accessed 14 September 2018].
Patterson, R. (2009) 'After a Decade of Hits, The Matrix Are Still Going Strong', *Music World Magazine*.
Peters, G. (2009) *The Philosophy of Improvisation.* Chicago, University of Chicago Press.
Peterson, R. and Kern, R. (1996) 'Changing Highbrow Taste: From Snob to Omnivore', *American Sociological Review* 61 (5), pp. 900–990.
Peterson, R. and Simkus, A. (1992) 'How Musical Taste Groups Mark Occupational Status Groups', in Lamont, M. and Fournier, M. (eds) *Cultivating Differences: Symbolic Boundaries and the Making of Inequality,* Chicago, University of Chicago Press.
Plato (1931) *The Dialogues of Plato, Volume IV*. Trans. B. Jowett. London, Oxford University Press.
Plotinus (1991) *The Enneads*. Trans. S. MacKenna. London and New York, Penguin Books.
Plunkett, J. (2012) 'Billy Bragg: "Education Reforms Risk Stifling Creativity"' [Online]. *The Guardian*. Available from www.guardian.co.uk/music/2012/nov/12/billy-bragg-education-reforms-stifle-creativity [Accessed 13 November 2012].
Powers, A. (2009) 'Pop Notes on the Decade: Authenticity Takes a Holiday' [Online]. Pop & Hiss: The LA Times Music Blog. Available from: http://latimesblogs.latimes.com/music_blog/2009/12/pop-music-notes-on-the-decade-authenticity-takes-a-holiday.html [Accessed 15 January 2013].
Prochak, T. (2001) *How to Remix*. London, Sanctuary Publishing Ltd.
Ranciere, J. (2011) *The Emancipated Spectator.* London and Brooklyn, Verso.
Realpe, A. and Wallace L. (2010) *What is Co-production?* London, The Health Foundation.
Rhodes, M. (1961) *An Analysis of Creativity. The Phi Delta Kappan,* 42 (7), pp.305–310.
Robjohns, H. (1999) 'Otari RADAR II' [Online]. Available from https://www.soundonsound.com/reviews/otari-radar-ii [Accessed 25 August 2020].
Robson, P. (2017) 'The Songs Remain the Same', *The Guardian*, 14 April 2017, section 2 pp. 4–7.
Rovane, C. (1998) *The Bounds of Agency: An Essay in Revisionary Metaphysics*. Princeton, Princeton University Press.
Ryle, A. and Fawkes, L. (2007) 'Multiplicity of Selves and Others: Cognitive Analytic Therapy', *Journal of Clinical Psychology,* 63 (2), pp.165–174.
Savona, A. (ed.) (2005) *Console Confessions: The Great Music Producers in Their Own Words.* San Francisco, CA, Backbeat Books.
Sawyer, K. (2007) *Group Genius: The Creative Power of Collaboration*. New York, Basic Books.
Sayer, A. (2005) *The Moral Significance of Class*. Cambridge, Cambridge University Press.

SOURDIEZELDVD (2009) 'Hell Rell by Your Side' [Online] Available from www.youtube.com/watch?v=zC0KADyUTks [Accessed 25 Aug 2020].

Stirner, M. (2014) *The Ego and His Own*. New York, Verso.

Seddon, F. (2004) 'Empathic Creativity: The Product of Empathic Attunement' in Miell, D. and Littleton, K. (eds) *Collaborative Creativity: Contemporary Perspectives*. London, Free Association Books, pp. 65–78.

Sharma, L. (2014) 'Recording with SAW' [Online]. Available from https://lessharma.com/recording-with-stock-aitken-waterman-saw/ [Accessed 22 July 2019].

Shen, M. (2005) 'What's Korn Got in Common with Avril, Britney and Mariah', *New York Post*, 27 November 2005.

Shuker, R. (2017) *Popular Music: The Key Concepts*, 4th ed. London, Routledge.

Skills for Care (2018) *Co-production in Mental Health: Not Just Another Guide*. London, Skills for Care.

Slay, J. and Stephens, L. (2013) *Co-production in Mental Health: A Literature Review*. London, New Economics Foundation.

Small, C. (1998) *Musicking: The Meanings of Performing and Listening*. Middletown, Wesleyan University Press.

Smith, G.F. (1998) 'Idea-Generation Techniques: A Formulary of Active Ingredients', *The Journal of Creative Behavior,* 32 (2), pp.107–134.

Smith, M.K. (2020) 'Pierre Bourdieu on Education: Habitus, Capital, and Field. Reproduction in the Practice of Education' [Online] Available from https://infed.org/mobi/pierre-bourdieu-habitus-capital-and-field-exploring-reproduction-in-the-practice-of-education/ [Accessed 19 November 2020].

Social Care Institute for Excellence (2015) *Guide 51: Co-production in Social Care: What It Is and How to Do It* [Online]. Available from www.scie.org.uk/publications/guides/guide51/ [Accessed 21 October 2020].

Softtube. (2010) 'Softube Visits The Matrix in the Making of BC Jean's Single "Just a Guy"' [YouTube 20 August 2010].

Spectropop (N.D.) *Curt Boettcher* [Online]. Available from www.spectropop.com/hsoftcurtb1.html [Accessed 23 February 2018].

Spinoza, B. (1996) *Ethics*. London, Penguin Books.

Staged. (2020) [TV Programme] Directed by Simon Evans. London, BBC One.

Sternberg, R.J. (1999) 'A Propulsion Model of Types of Creative Contributions', *Review of General Psychology,* 3 (2), pp.83–100.

Sternberg, R.J. (2006) 'The Nature of Creativity', *Creativity Research Journal,* 18, (1), pp. 87–98.

Stewart, I. (1987) *From Here to Eternity*. Oxford and New York, OUP.

Stock, M. (2004) The Hit Factory. London, New Holland Publishers.

Strauss, N. (2002) *The Pop Life: Given Up, A Dream Returns to Life*. The New York Times 20 September 2003.

Strachan, R. (2017) *Sonic Technologies: Popular Music, Digital Culture and the Creative Process*. New York, Bloomsbury Publishing.

Tamm, E. (1995). *Brian Eno: His Music and the Vertical Colour of Sound*. Boston, Da Capo Press.

Taylor, T.D. (1997) Global *Pop: World Music, World Markets*. New York, Routledge.

Thaler, R. and Sunstein, C. (2008) *Nudge: Improving Decisions About Health, Wealth and Happiness*. London and New York, Penguin.

The Frames (2005) *Live at Park West*. Internet Archive. Available from http://archive.org/details/frames2005-10-22.flac16 [Accessed 10 November 2021].

The Frames (2003) *Set List*. Plateau Records.

The Matrix.com (2018) 'Our Passion Is Music' [Online]. Available at http://thematrixmusic.com [Accessed 28 September 2018].

TheRealHipHopChannel (2010) 'ARAABMUZIK ... By Your Side Freestyle – Vya-B' [2010/CDQ] [Online]. Available from www.youtube.com/watch?v=2EuI6348NBo [Accessed 25 Aug 2020].

Thomas, R. (2019) 'In Pictures: Waiting for Curtain Up' [Online]. Available from www.bbc.com/news/entertainment-arts-49495357 [Accessed 8 September 2019].

Thompson, P. (2019) Creativity in the *R*ecording *S*tudio. New York, Palgrave Macmillan.

Tuckman, B.W. (1965) 'Developmental Sequence in Small Groups', *Psychological Bulletin*, 63(6), pp.384–399.

Turner, A. (2014) *BRIT Awards, acceptance speech.* 19 February 2014. (ITV Broadcasting)
Vallerand, R.J., Blanchard, C., Mageau, G.A., Koestner, R., Ratelle, C., Léonard, M., Gagné, M., and Marsolais, J. (2003) 'Les Passions de l'Âme', *Journal of Personality and Social Psychology,* 85 (4), pp.756–767.
Wacquant, L. (2014) 'Homines in Extremis: What Fighting Scholars Teach Us About Habitus', *Body & Society,* 20 (2), pp.3–17.
Wacquant, L. (2016) 'A Concise Genealogy and Anatomy of Habitus', *The Sociological Review (Keele),* 64 (1), pp.64–72.
Waterman, P. (2000) *i wish i was me: The Autobiography.* London, Virgin Publishing Ltd.
Websman, A. (2013) 'Lauren Christy: Quintessence of a Woman', Miroir Magazine, 20 December 2013.
Weinstein, D. (1997) *The History of Rock's Pasts through Rock Covers.* In Swiss, T.; Sloop J. and Herman, A. (eds) *Mapping the Beat.* Oxford, Blackwell Publishing Ltd.
Weisethaunet, H. and Lindberg, U. (2010) 'Authenticity Revisited: The Rock Critic and the Changing Real', *Popular Music and Society*, 33 (4), pp.465–485.
Williams, A. (2012) 'I'm Not Hearing What You're Hearing: The Conflict and Connection of Headphone Mixes and Multiple Audioscapes', in Frith, S. and Zagorski-Thomas, S, (eds) *The Art of Record Production: An Introductory Reader for a New Academic Field.* Surrey and Burlington, VT, Ashgate, pp.113–128.
Wilsmore, R. (2004) *Selection 44*, for Orchestra *f.p.* Viva The Orchestra of the East Midlands, Cond. Nicholas Cox.
Wilsmore, R. (2010) 'The Demonic and the Divine: Unfixing Replication in the Phenomenology of Sampling', *The Journal of Music, Technology & Education*, 3 (1), pp.5–16.
Wilsmore, R. and Johnson, C. (2017) 'The Mix Is. The Mix Is Not', *Perspectives on Music Production: Mixing*. London and New York, Routledge.
Wikipedia (2018) 'Deterritorialization'. Entry in Wikipedia [Online]. Available from https://en.wikipedia.org/wiki/Deterritorialization [Accessed 25 January 2018].
Witkin, R.W. (2003) *Adorno on Popular Culture.* London, Routledge.
Wittgenstein, L. (1953) *Philosophical Investigations*. Trans. G. E. Anscombe. New York, Macmillan Publishing.
Yamaha (2018) *The Matrix Biogs* [Online]. Available from yamaha.com [Accessed 12 October 2018].
Zak, A. (2001) *The Poetics of Rock: Cutting Tracks, Making Records*. London, University of California Press.

Selected Discography

Aphex Twin. (2003) *26 Remixes for Cash.* Warp Records.
Bowie, D. and Eno, B. (1977) *Low*. RCA Records.
Burn, P. *et al.* (1984) 'You Spin Me Round (Like a Record)'. Epic Records.
Chemical Brothers (2002) Come with Us, Astralworks.
Christina Aguilera (2000) *My Kind of Christmas.* RCA Records.
Christy, L. (1993) *Lauren Christy*. Mercury Records.
Christy, L. (1997) *Breed.* Mercury Records.
Christy, L., Spock, S., Edwards, G., and Lavigne, A. (2002) 'Complicated'. Arista Records.
Christy, L., Spock, S., Edwards, G., and Lavigne, A. (2002) 'Sk8ter Boi'. Arista Records.
Collins, P. (1980) 'In the Air Tonight'. Virgin Records, Atlantic Records.
Dee Gees (Foo Fighters). (2021) *Hail Satin.* RCA Records.
DJ Danger Mouse. (2004) *The Grey Album.*
Dylan, Bob (1997) *Time out of Mind.* Colombia Records.
Gabriel, P. (1980) *Peter Gabriel.* Charisma Records, Mercury Records.
Grant, E., Nowels, R., and Menzies, K. (2017) 'Get Free'. Polydor Records, Interscope Records.
Hammond, A., and Hazlewood, M. (1974) 'The Air That I Breathe'. Polydor Records, Epic Records.
Hammond, A., Hazlewood, M., and Radiohead. (1992) 'Creep', Parlophone, Records, EMI Records.
Harding, P. (2008) *The Story of Beginners.* WB Records.

Holland, B., Dozier, L., and Holland, E. (1965) 'Stop! in the Name of Love'. Motown Records.
Horn, T., Downes, G., and Woolley, B. (1979) 'Clean Clean'. Epic Records.
Horn, T., Downes, G., and Woolley, B. (1979) 'Video Killed the Radio Star'. Epic Records.
Horn, T., Downes, G., and Woolley, B. (1980) 'Clean Clean'. Island Records.
Horn, T., Downes, G., and Woolley, B. (1980) 'Video Killed the Radio Star'. Island Records.
Jones, G. (2008) *Hurricane.* Wall of Sound, PIAS.
Jones, G. and Woolley, B. (1986) 'Party Girl'. Manhattan Records.
Jones, G., Woolley, B., and Van Eyck, M. (2010) 'Love You to Life'. Wall of Sound.
Killing Joke. (1980) *Killing Joke.* E.G. Polydor Records.
Korn. (2005) *See You on the Other Side.* EMI Records, Virgin Records.
Lavigne, A. (2002) *Let Go.* Arista Records.
Martyn, J. (1973) *Solid Air* [ILPS 9226]. UK, Island Records.
Mitchell, J. (1972) *Blue* [K 44128]. UK, Reprise Records.
Mostly Autumn. (2007) *Heartful of Sky.* Mostly Autumn Records [AUT933].
Mortimer, T., Kean, R., and Hawken, D. (1994) 'Stay Another Day'. London Records.
Richard, C. (2011) *Soulicious.* Parlaphone Records.
Shone, L., Matthias, J., Joyce, M., Owen, S., Parsons, J., Barker, S., and Hirst, C. (1982) 'Sign of the Times.' Stiff Records.
Squarepusher (1999) *Selection Sixteen.* Warp Records.
Stock, M., Aitken, M., and Waterman, P. (1985) 'Say I'm Your Number One'. Supreme Records.
Stock, M., Aitken, M., and Waterman, P. (1986) Showing Out (Get Fresh at the Weekend) System'. Blow up. RCA. Atlantic.
Stock, M., Aitken, M., and Waterman, P. (1987) 'I Should Be So Lucky'. PWL Records.
Stock, M., Aitken, M., and Waterman, P. (1987) 'Respectable' Supreme.
Stock, M., Aitken, M., and Waterman, P. (1987) 'Never Gonna Give You Up'. RCA Records, PWL Records.
Strummer, J. and Jones, M. (1978) *(White Man) in Hammersmith Palais.* CBS Records.
Talking Heads (1978) *More Songs About Buildings and Food.* Sire Records.
Talking Heads (1979) *Fear of Music.* Sire Records.
Taylor, R.D., Wilson, F., Sawyer, P., and Richards, D. (1967) 'Love Child'. Motown Records.
The And (N.D.) *The Song of a Thousand Songs.* Mayditup Records.
The Frames (2002) *Breadcrumb Trail.* Plateau Records.
The Buggles (1980) *Living in the Plastic Age.* Island Records.
The Camera Club (1979) *English Garden.* Epic Records.
Unknown (N.D.) 'This Song Sounds Like Another Song'. Mayditup Records.
U2 (1987) *The Joshua Tree.* Island Records.
Various Artists (2019) *Smoke Rainbows – Music Minds Matter* [HMUK001]. London, Monks Road Records.
Whitfield, N. and Strong, B. (1968) 'I Heard It Through the Grape Vine'. Tamla Records.
Woolley, B. Darlow, S., Lipson, S., and Horn, T. (1985) 'Slave to the Rhythm'. Island Records, Manhattan Records.

Index

Note: References to figures and photographs appear in *italic* type; those in **bold** type refer to tables.

Printed in the United States
by Baker & Taylor Publisher Services